INSTRUMENTATION IN THE CHEMICAL AND PETROLEUM INDUSTRIES, VOL. 13

Programmed by:
ISA's Chemical and Petroleum Industries Division

Proceedings from the 1977
Spring Industry Oriented Conference
Anaheim, California

INSTRUMENTATION IN THE CHEMICAL AND PETROLEUM INDUSTRIES, VOL. 13

Proceedings of the 14th Annual ISA Chemical
and Petroleum Instrumentation Symposium

May 2-5, 1977, Anaheim, California

Instrumenting and Controlling
Centrifugal Compressors

Edited by

R. C. Waggoner

*University of Missouri—Rolla
Rolla, Missouri*

INSTRUMENT SOCIETY OF AMERICA
Pittsburgh, Pennsylvania

PUBLICATION POLICY

Technical papers may not be reproduced in any form without written permission from the Instrument Society of America. The Society reserves the exclusive right of publication in its periodicals of all papers presented at the Annual ISA Conference, at ISA Symposia, and at meetings co-sponsored by ISA when the Society acts as publisher. Papers not selected for such publication will be released to authors upon request.

In any event, following oral presentation, other publications are urged to publish up to 300-word excerpts of ISA papers, provided credits are given to the author, the meeting, and the Society using the Society's name in full, rather than simply "ISA".

Reprints of articles in this publication are available on a custom printing basis at reasonable prices in quantities of 50 or more.

For further information concerning publications policy and reprint quotations, contact:

Publications Department
Instrument Society of America
400 Stanwix Street
Pittsburgh, Pa. 15222
Phone: (412) - 281-3171

Library of Congress Catalog Card Number 64-7505

ISBN 87664-363-2

Copyright © 1977

INSTRUMENT SOCIETY OF AMERICA
400 Stanwix Street
Pittsburgh, Pa. 15222

Printed in U.S.A.

DIRECTOR'S FOREWORD

1977 is the eighteenth (18th) consecutive year that the Chemical and Petroleum Industries Division of ISA has presented a spring symposium. The proceedings volume from this CHEMPID symposium has been assembled in keeping with the current interests of the chemical and petroleum industries to serve as a permanent record and a guide to today's process systems technology. It is intended to benefit those who continue to lead the way in the conception, design, and implementation of process control.

The program coordinator for these CHEMPID sessions has been Dr. Raymond C. Waggoner, University of Missouri—Rolla. Dr. Waggoner, our Education Chairman, has worked diligently to put together an informative group of papers.

K. L. Hopkins
Chempid Director

PROGRAM CHAIRMAN

The papers in this volume were programmed by the Chemical and Petroleum Industries Division and were presented at the ISA/77 Conference, May 2-5, in Anaheim, California. These papers implement the conference theme "Leading the Way in Control Systems" by presenting applications of advanced control techniques in the chemical and petroleum industry. Processes in these industries are inherently multi-variable and non-linear. Advanced control techniques are requisite if the most economical and effective operating levels are to be sustained. However, the complexities of the control algorithms virtually require digital computer implementation.

The fundamental concepts of advanced control techniques are presented to provide a background for the following papers. These advanced concepts, including feedforward control, multivariable control, optimal control, and digital control algorithms are developed and papers describe their application to actual processes, pilot plant scale equipment, or calibrated process simulations. A group of papers is specifically directed to chemical reactor control. The economic aspects of advanced control are then shown in the papers taken from the final two sessions.

This volume is presented to illustrate that advanced control concepts can be made practical in chemical and petroleum processes and to aid the reader in their implementation.

R. C. Waggoner
Program Coordinator

CONTENTS

SESSION: LEADING THE WAY IN CONTROL SYSTEMS
Chairman: R. C. Waggoner

IMPROVED CONTROL BY APPLICATION OF ADVANCED CONTROL
TECHNIQUES, R. K. Wood 1

SESSION: APPLICATION OF NEW AUTOMATIC CONTROL CONCEPTS IN CHEMICAL REACTORS
Chairman: F. R. Groves

A COMPARISON OF CONTROLLER ALGORITHMS AS APPLIED TO A STIRRED
TANK REACTOR, B. T. Condon, C. A. Smith 11

A SENSITIVITY ANALYSIS ON DAHLIN'S CONTROL ALGORITHM,
B. T. Condon, C. A. Smith 21

COMPARISON OF TUNING METHODS FOR TEMPERATURE CONTROL OF A
CHEMICAL REACTOR, J. Martin, Jr., A. B. Corripio, C. L. Smith 31

EMPIRICAL SECOND ORDER NONLINEAR PROCESS MODEL DEVELOPMENT
APPLICATION, C. A. Smith, F. R. Groves, Jr. 37

SESSION: ADVANCED CONTROL CONCEPTS
Chairman: R. A. Mollenkamp

EVALUATION OF VARIOUS CONTROL STRATEGIES FOR A POWER PLANT
FUEL OIL HEATING CONTROL SYSTEM, T. P. Davis, C. A. Smith 47

ADAPTIVE CONTROL THROUGH INSTRUMENTAL-VARIABLE ESTIMATION OF
DISCRETE MODEL PARAMETERS, A. T. Touchstone, A. B. Corripio 57

DEVELOPMENT OF MULTIVARIABLE CONTROL STRATEGIES FOR
DISTILLATION COLUMNS, C. O. Schwanke, T. F. Edgar, J. O. Hougen 65

APPLICATION OF A SIMPLE DISCRETE MULTIVARIABLE STRATEGY TO A
LABORATORY SCALE AUTOCLAVE, B. Sayers, C. F. Moore 79

SESSION: ADVANCED CONTROL SYSTEM DESIGN IN CHEMICAL AND PETROLEUM PLANTS
Chairman: D. M. Steelman

EVALUATION OF CONTROLLERS FOR DEAD-TIME PROCESSES, C. W. Ross 87

DIGITAL COMPUTER CONTROL OF AMMONIA PLANTS, J. S. Gruneisen 99

AN IMPROVED COMPUTER CONTROLLED PROCESS CHROMATOGRAPH
SYSTEM, M. J. Hausner 103

HIGH-SPEED DATA ACQUISITION OF EVENTS ON A VOLATILE PROCESS,
G. E. Pease, Jr., L. C. Boxhorn 107

AUTHOR INDEX 113

APPENDIX 114

ACKNOWLEDGMENTS 115

IMPROVED CONTROL BY APPLICATION

OF ADVANCED CONTROL TECHNIQUES

R.K. Wood
Department of Chemical Engineering
University of Alberta
Edmonton, Alberta, Canada

ABSTRACT

Sophisticated computer-aided design procedures leading to specified optimal operating conditions for process units, are now in use in many design groups. Accompanying this change in design techniques is an increased awareness of the necessity of developing control system strategies, based on advanced control techniques, to achieve the desired optimal operation.

Of the advanced techniques leading to improved control performance, feedforward control has gained the most acceptance by industry. Combined feedforward-feedback control will be briefly discussed. Further improvements in control system design have resulted by considering the multivariable nature of the process system to be controlled. Although the subject of much attention, design based on linear state space models has not been too successful due to the difficulty of characterizing process systems by such a model. Consequently the discussion will be restricted to multivariable frequency domain design methods. Non-interacting/decoupling control and design by the characteristic loci technique will be considered in this review.

INTRODUCTION

Rapid advances in computer technology have now made computer-aided design techniques standard practice in many engineering departments. As a consequence of this development, frequently the designer will specify a unit on some optimal design basis that is very dependent on operating conditions. In order to derive maximum economic benefit from the improved design, it logically follows that this can only be achieved by improved control of the system. In conjunction with the use of computers for design has come the utilization of computers for process control. Although, initially the control algorithms were simply replacements for conventional single variable feedback PID controllers, it was soon realized that to achieve the specified operating conditions that more advanced control techniques would be required. Probably the first advanced control technique to gain acceptance in the process industries was that of feedforward control, and then not until the early 1960's. However as Buckley (1) correctly points out, the idea of feedforward control was not a new control concept at all. Feedforward control had already been used for many years for manipulating feedwater rate to improve level control in boiler drums. The 1960 decade also saw the development of what came to be known as modern control theory, namely the body of control theory predicated on describing system behaviour in the time domain by means of a linear state space model. A typical model formulation would be of the following form:

$$\underline{\dot{x}} = \underline{\underline{A}}\,\underline{x} + \underline{\underline{B}}\,\underline{u} + \underline{\underline{D}}\,\underline{d} \qquad (1)$$

$$\underline{y} = \underline{\underline{C}}\,\underline{x} \qquad (2)$$

where $\underline{x}$ = vector of "n" state variables
$\underline{u}$ = vector of "m" control inputs
$\underline{d}$ = vector of "p" disturbance inputs
$\underline{y}$ = vector of "q" output variables
$\underline{\underline{A}}, \underline{\underline{B}}, \underline{\underline{D}}, \underline{\underline{C}}$, = matrices of appropriate order

Hundreds of articles have appeared which develop the theory necessary for establishing control laws for specific types of models using a wide variety of performance indices. However, much of the work is not directly applicable to the regulatory control problems of the process industries. The reader interested in gaining an appreciation of the design approach involved when a system is characterized by a state space model should consult the case study of Fisher and Seborg (2). The authors present experimental results from applying several different multivariable control system strategies to a pilot plant evaporator that could adequately be characterized by a model of the form expressed by Equations (1) and (2).

Unfortunately, the nonlinear characteristics of most process systems are such that it is very difficult to describe their behaviour by such a linear model. As noted by Rijnsdorp and Seborg (3), few applications of control schemes based on such a model have been adopted by the process industries. In contrast, the authors noted that non-interacting or decoupling control schemes, based on a transfer function representation, had gained wide acceptance by industry. Such a control scheme is but one approach to designing a control system using multivariable frequency domain techniques pioneered by Rosenbrock, MacFarlane and co-workers at the University of Manchester Institute of Science and Technology in England. Research has demonstrated that the single variable concepts of Bode and Nyquist can be extended to developing control strategies for multivariable systems. The design techniques that have been developed are known as the inverse Nyquist array (4,5,6); characteristic loci (7,8,9,10); direct

Nyquist array (11,12,13) and commutative controller (14,15,16). Another class of design techniques are those based on the sequential return difference method originally developed by Mayne (17,18,19) and subsequently modified by Owens (20,21). A recent evaluation of the characteristic loci and both the inverse and direct Nyquist array design techniques has been performed by Kuon (22).

In this review, the accepted technique of combined feedforward-feedback control will first be discussed. This material is included to emphasize that implementation of feedforward action is simple and worthy of consideration for improving the control of any process system. The remainder of the discussion will be directed to reviewing the non-interacting/decoupling control and characteristic loci multivariable frequency domain design techniques. It is hoped that by drawing attention to such multivariable techniques that designers of control systems will not be reluctant (23) to consider such techniques.

COMBINED FEEDFORWARD-FEEDBACK CONTROL

Applications of feedforward control action combined with conventional feedback control did not become routine until the installation of process control computers. With the advent of the process control computer, implementation no longer meant that a special pneumatic or electronic unit (24,25) had to be constructed. Despite the ease with which a feedforward control loop can now be implemented, using standard electronic feedforward control units or by computer, it is surprising that more combined feedforward-feedback systems are not utilized. Implementation of a feedforward control loop can lead to a substantial cost saving (26) for a small capital investment.

The block diagram of a combined feedforward-feedback control system is shown in Figure 1. The control variable can be expressed in terms of the disturbance, $D(s)$ and the set point, $R(s)$ as:

$$C(s) = \left\{\frac{G_c G_v G_p}{1 + G_{OL}}\right\} R(s) + \left\{\frac{G_L}{1 + G_{OL}}\right\} D(s) + \left\{\frac{G_{MD} G_{FF} G_v G_p}{1 + G_{OL}}\right\} D(s) \quad (3)$$

where $G_{OL} = G_c G_v G_p G_{MC}$. Writing the expression for $C(s)$ in this form, serves to emphasize that the stability of the control system is unchanged by the addition of feedforward control action and furthermore if there is no feedforward control, $G_{FF} = 0$, the expression is that for a conventional feedback control system. Considering the case of regulatory control ($R(s) \equiv 0$), allows Equation (1) to be written as

$$C(s) = \left\{\frac{G_L + G_{MD} G_{FF} G_v G_p}{1 + G_{OL}}\right\} D(s) \quad (4)$$

Since the control objective is no change in the control variable for load upsets, it follows from Equation (4) that to satisfy this condition it is necessary that $G_L + G_{MD} G_{FF} G_v G_p = 0$, or rearranging

$$G_{FF} = \frac{-G_L}{G_{MD} G_v G_p} \quad (5)$$

Equation (5) is the specified form of the feedforward controller, G_{FF}. It is particularly interesting to note the form that the feedforward controller takes under certain circumstances. If the load and process transfer functions, G_L and G_p respectively, have the same dynamics and the valve, G_v and measurement device, G_{MD} dynamics can be neglected then Equation (5) becomes

$$G_{FF} = \frac{-K_L}{K_{MD} K_v K_p} \quad (6)$$

In this particular case, the required feedforward controller is simply a gain device, and readily implemented by cheap conventional instrumentation or even by the measurement device itself. For the case where the dynamics of the load and process transfer functions are the same, the dynamics of the valve represented by $K_v/(\tau_v s+1)$ and the dynamics of the measurement device neglected then Equation (5) can be written as

$$G_{FF} = \frac{-K_L(\tau_v s+1)}{K_{MD} K_v K_p} \quad (7)$$

This form of the feedforward controller is simply a PD controller! More typically, the load and process transfer function dynamics are not the same and the dynamics of the valve and measurement device can be neglected giving the required form of the controller as

$$G_{FF} = \frac{-G_L(s)}{K_{MD} K_v G_p(s)} \quad (8)$$

Obviously, it would be a simple matter to implement the required controller action by means of a digital computer. However in actual practice despite the fact that $G_L(s)$ and $G_p(s)$ may be of high order, experience has shown (26) that it is generally adequate to consider the feedforward controller as

$$G_{FF} = \frac{-K(1 + \tau_1 s)}{(\tau_2 s + 1)} \quad (9)$$

and employ on-line tuning of the parameters. Although this type of lead-lag unit, proposed several years ago by Shinskey (27), has been installed for many applications, experience has shown that satisfactory control can be achieved without employing lead action. This experience is substantiated by tests on a pilot scale distillation column by Wood and Pacey (28) who found that satisfactory control could be achieved using only a tuned first order lag or time delay element. It is important to realize that the choice of the form of feedforward controller can significantly influence system control behaviour. For instance in the paper by Davis and Smith (29), in this symposium, they considered only gain feedforward action. As can be seen in their Figure 5, without dynamic compensation, the controlled variable was initially driven below the set point. It would be interesting to compare the control behaviour of the combined feedforward-feedback control scheme, using a dynamic feedforward controller, with that of the dual control loop scheme (in which the effect of the disturbance is minimized by the temperature loop).

MULTIVARIABLE FREQUENCY DOMAIN DESIGN

Frequently the design of a process control system will involve the control of more than a single output variable by manipulating more than one input variable. Such a system is multivariable, so a control system designed using conventional single variable theory may not yield satisfactory performance. The control performance using multiple single variable feedback loops will depend on the extent of interaction between the input and output variables. In order to establish whether such an approach will be satisfactory the degree of interaction can be checked using the procedure suggested by Bristol (30). Since the development of control strategies based on a state space model have had very limited success, the discussion here will focus on two design techniques for systems characterized by a transfer function representation.

A) NON-INTERACTING/DECOUPLING DESIGN TECHNIQUE

This design technique, directed at reducing the multivariable feedback design problem to one of conventional single loop design, was first proposed by Boksenbom and Hood (31). In this approach, decoupling controllers that render the system completely non-interactive are first designed, and then conventional single variable design methods are employed to design feedback controllers. Despite the possible difficulties (32,33) with such a procedure, it has been successfully employed to pilot scale and industrial process units (3,34,35, 36). Although in theory the concept could be extended to high order dynamic systems, reported applications have been concerned only with control of two variables by manipulation of two input variables.

The design procedure will now be outlined for the case of a 2 x 2 plant transfer function matrix, $\underline{\underline{G}}_p(s)$. The block diagram of such a system is shown in Figure 2 where $U_1(s)$, $U_2(s)$ are the input variables; $C_1(s)$, $C_2(s)$ the output (controlled) variables and $D(s)$ the load disturbance. In vector-matrix form this may be expressed as

$$\underline{C}(s) = \underline{\underline{G}}_p(s) \underline{U}(s) + \underline{G}_L(s) D(s) \quad (10)$$

where

$$\underline{C}(s) = \begin{bmatrix} C_1(s) \\ C_2(s) \end{bmatrix} \quad \underline{U}(s) = \begin{bmatrix} U_1(s) \\ U_2(s) \end{bmatrix} \quad \underline{G}_L(s) = \begin{bmatrix} G_{L1} \\ G_{L2} \end{bmatrix}$$

$$\underline{\underline{G}}_p(s) = \begin{bmatrix} G_{11} & G_{12} \\ G_{21} & G_{22} \end{bmatrix}$$

Now, if a feedback controller and measurement device are included for each of the controlled variables, that is G_{C11}, H_1, G_{C22}, H_2 as well as two additional controllers, G_{C12}, G_{C21}, then the system can be represented as shown by the block diagram in Figure 3. (Note: The control valve transfer functions are considered to be included in the plant transfer function matrix). The two controllers, G_{C12} and G_{C21} are generally designated as decoupling controllers or compensators. Defining

$$\underline{R}(s) = \begin{bmatrix} R_1(s) \\ R_2(s) \end{bmatrix} \quad \underline{\underline{H}}(s) = \begin{bmatrix} H_1 & 0 \\ 0 & H_2 \end{bmatrix}$$

$$\underline{\underline{G}}_C(s) = \begin{bmatrix} G_{C11} & G_{C12} \\ G_{C21} & G_{C22} \end{bmatrix}$$

allows the block diagram to be redrawn in general multivariable feedback form as shown in Figure 4. With the block diagram arranged to the same form as that of a conventional single variable feedback control system, the following expression logically follows:

$$\underline{C}(s) = [\underline{\underline{I}} + \underline{\underline{G}}_p(s)\underline{\underline{G}}_C(s)\underline{\underline{H}}(s)]^{-1} \{\underline{\underline{G}}_C(s)\underline{\underline{G}}_p(s)\underline{R}(s) + \underline{G}_L(s)D(s)\} \quad (11)$$

where $\underline{\underline{I}}$ = identity matrix.
Defining $\underline{\underline{Q}}(s) = \underline{\underline{G}}_p(s)\underline{\underline{G}}_C(s)$, as the <u>open loop transfer function matrix</u> allows Equation (11) to be expressed as

$$\underline{C}(s) = [\underline{\underline{I}} + \underline{\underline{Q}}(s)\underline{\underline{H}}(s)]^{-1} \{\underline{\underline{Q}}(s)\underline{R}(s) + \underline{G}_L(s)D(s)\} \quad (12)$$

Defining the <u>closed loop transfer function matrix</u> as

$$\underline{\underline{P}}(s) = [\underline{\underline{I}} + \underline{\underline{Q}}(s)\underline{\underline{H}}(s)]^{-1} \underline{\underline{Q}}(s)$$

allows Equation (12) to be written as

$$\underline{C}(s) = \underline{\underline{P}}(s)\underline{R}(s) + \underline{\underline{P}}(s)\underline{\underline{Q}}(s)^{-1} \underline{G}_L(s)D(s) \quad (13)$$

It now follows that if $\underline{\underline{Q}}(s)\underline{\underline{H}}(s)$ can be diagonalized, then $[\underline{\underline{I}} + \underline{\underline{Q}}(s)\underline{\underline{H}}(s)]^{-1}$ will be diagonal since $\underline{\underline{H}}(s)$ is already diagonal. Therefore the closed loop transfer function matrix, $\underline{\underline{P}}(s)$ will be diagonal which means that there will be no interaction between $\underline{C}(s)$ and $\underline{R}(s)$ (or $D(s)$). From Figure 4, it can be shown that

$$\underline{\underline{Q}}(s)\underline{\underline{H}}(s) = \begin{bmatrix} \alpha_{11} & \alpha_{12} \\ \alpha_{21} & \alpha_{22} \end{bmatrix}$$

where

$$\alpha_{11} = G_{11}G_{C11}H_1 \left[1 + \frac{G_{12}G_{C21}}{G_{11}G_{C11}}\right] \quad (14)$$

$$\alpha_{12} = \{G_{11}G_{C12} + G_{12}G_{C22}\} H_2 \quad (15)$$

$$\alpha_{21} = \{G_{21}G_{C11} + G_{22}G_{C21}\} H_1 \quad (16)$$

$$\alpha_{22} = G_{22}G_{C22}H_2 \left[1 + \frac{G_{21}G_{C12}}{G_{22}G_{C22}}\right] \quad (17)$$

The requirement for diagonalization means that the off-diagonal terms must be zero, that is $\alpha_{12} = 0$, $\alpha_{21} = 0$. Setting the off-diagonal entries to zero gives the required form of the decoupling controllers as

$$G_{C12} = -\frac{G_{12}G_{C22}}{G_{11}} \quad (18)$$

$$G_{C21} = -\frac{G_{21}G_{C11}}{G_{22}} \quad (19)$$

Substituting for the decoupling controllers, as given by Equations (18) and (19), in Equations (14) and (17) gives

$$\alpha_{11} = G_{C11}G_{11}H_1 \left[1 - \frac{G_{12}G_{21}}{G_{11}G_{22}}\right] = G_{C11}\hat{G}_{11} \quad (20)$$

$$\alpha_{22} = G_{C22}G_{22}H_2 \left[1 - \frac{G_{12}G_{21}}{G_{11}G_{22}}\right] = G_{C22}\hat{G}_{22} \quad (21)$$

which in turn allows the closed loop transfer function matrix, $\underline{\underline{P}}(s)$ to be written as

$$\underline{\underline{P}}(s) = \begin{bmatrix} \frac{\alpha_{11}H_1^{-1}}{1 + \alpha_{11}} & 0 \\ 0 & \frac{\alpha_{22}H_2^{-1}}{1 + \alpha_{22}} \end{bmatrix}$$

Expressing α_{11} and α_{12} in Equations (20) and (21) in terms of $\hat{G}_{11}$ and $\hat{G}_{22}$ is done to emphasize that these groups are calculated from plant data. All that remains is selection of the controllers G_{C11} and G_{C22} which can proceed on the basis of two separate single variable systems since the interaction has been eliminated. Once the desired feedback controllers have been determined the required form of the decoupling controllers can be calculated from Equations (18) and (19).

B) CHARACTERISTIC LOCI DESIGN TECHNIQUE

An alternate approach to the design of a completely non-interacting/decoupled control system is simply to employ a design procedure that minimizes but not totally eliminates interaction. Design techniques that employ such an approach strive to achieve diagonal dominance. A control system designed on this basis may actually provide better control than is possible with the non-interacting system. This will depend upon the reliability/variability of the transfer function parameters that determine the parameters of the decoupling controllers.

A comprehensive survey of the existing multivariable frequency domain procedures has been presented in the excellent five part review of MacFarlane (37). Since the characteristic locus method as well as the non-interacting design approach has been employed in the paper of Schwanke et al. (38), to be presented in this symposium, this discussion will deal only with this technique.

In order to restrict the length of this review, some fundamental definitions, concepts, conditions and/or requirements will simply be stated without proof. Most of the fundamental theoretical concepts have been presented by Belletrutti (8), MacFarlane (9) and MacFarlane and Belletrutti (7,10). Since s is a complex variable, then for every specific value of s (over the domain of definition, $\mathbb{C}$) it follows that an m x m matrix function of a complex variable $\underline{G}(s)$ is a matrix with complex entries. Thus it has a set of eigenvalues $\{g_i(s): i = 1, 2, \ldots m\}$ such that

$$g_i(s) \in \mathbb{C} \quad i = 1, 2, \ldots m$$

and corresponding sets of eigenvectors

$$\underline{d}_i(s) \in \mathbb{C}^n \quad i = 1, 2, \ldots m$$

(Note: A vector function of a complex variable, say $\gamma(s)$, is a mapping $\gamma(s): \mathbb{C} \to \mathbb{C}^n$ from the set of complex numbers $\mathbb{C}$ to the set of complex vectors $\mathbb{C}^n$). This notation means that the eigenvalues of a matrix function of a complex variable are functions of a complex variable, and the corresponding eigenvectors are vector functions of a complex variable. The eigenvalues, $g_i(s)$ of $\underline{G}(s)$ are designated as characteristic transfer functions while the corresponding eigenvector, $\underline{d}_i(s)$ is called the characteristic direction vector.

Also the set of loci in the complex plane obtained by evaluating a characteristic transfer function, $\underline{G}(s)$ along the standard Nyquist contour is known as the set of system characteristic loci and is denoted as $\{g_i(j\omega)\}$. The design considerations involve consideration of the open-loop transfer function matrix, $\underline{Q}(s)$ expressed in dyadic form as

$$\underline{Q}(s) = \sum_{i=1}^{m} q_i(s) \underline{w}_i(s) \underline{v}_i^T(s) \quad (22)$$

where $q_i(s)$, $w_i(s)$ and $v_i(s)$ are the characteristic transfer functions, characteristic direction vectors and reciprocal characteristic direction vectors of $\underline{Q}(s)$. The corresponding dyadic expansion for the closed-loop system can be written as

$$\underline{R}(s) = \sum_{i=1}^{m} \left|\frac{q_i(s)}{1 + q_i(s)}\right| w_i(s) v_i^T(s) \quad (23)$$

The design procedure based upon these and related theoretical concepts, is a generalization of the classical frequency domain approach, involving the conflicting objectives of stability, integrity, non-interaction and accuracy. This is accomplished by attaining required closed-loop stability and performance specifications by appropriate manipulations of sets of open loop characteristic loci and characteristic directions. Simplification of the task is achieved by letting the feedback matrix $\underline{H}(s) = \underline{I}$, since it can then be shown that if $\{q_i(j\omega)\}$ is the set of characteristic loci of the open loop system $\underline{Q}(s)$, then the set of characteristic loci belonging to the closed loop system $\underline{R}(s)$ is simply

$$\{\frac{q_i(j\omega)}{1 + q_i(j\omega)}\}$$

Furthermore the set of characteristic directions for both the open-loop and closed-loop systems are the same, namely $\{w_i(j\omega)\}$. Thus, the design effort is concerned with synthesizing the controller $\underline{G}_c(s)$ (cf. Figure 4), which is considered to be square as is the plant, $\underline{G}_p(s)$.

Design in the vector frequency response approach requires that the controller

i) modify the phases of appropriate sets of characteristic loci in order to achieve acceptable stability and integrity results.

ii) align the characteristic directions at high frequencies and balance the gains of the characteristic loci at low frequencies in order to achieve acceptable interaction.

iii) inject gain to improve overall performance.

Clearly, the controller must satisfy many objectives simultaneously. Thus a controller structure formed as a cascaded combination of several sub-controllers, $\underline{G}_{c_i}(s)$ so that

$$\underline{G}_c(s) = \prod_{i=1}^{\beta} \underline{G}_{c_i}(s)$$

is employed in which each of the $\underline{G}_{c_i}(s)$ achieves only part of the overall design objectives. Obviously each of the $\underline{G}_{c_i}(s)$ must be simple and if possible only contain constant factors. Specific restrictions on the $\underline{G}_{c_i}(s)$ are:

i) all dynamical elements must be rational functions in "s".

ii) det $\underline{G}_{c_i}(s)$ must be identically non-singular.

iii) poles of $\underline{G}_{c_i}(s)$ must lie in the open left-half plane.

iv) det $\underline{G}_{c_i}(s)$ must not have any right-half plane zeros (to prevent non-minimum phase difficulties).

Many different types of sub-controllers which provide for certain manipulations of system characteristic loci and characteristic directions have been developed. Some of the more useful (7,10) types of sub-controllers are those that provide for

a) Elementary transformation

i) $\underline{G}_{c_i}(s) = \text{diag}\{1,1, \ldots g_{jj}(s), \ldots 1\}$

ii)
$$\underline{G}_{c_i}(s) = \begin{bmatrix} 1 & 0 & 0 & 0 & \cdots & 0 & 0 \\ 0 & 1 & 0 & 0 & & . & 0 \\ 0 & 0 & 1 & g_{jk}(s) & & . & . \\ . & . & & 0 & & . & . \\ . & . & . & & & 1 & . \\ 0 & . & 0 & \cdots\cdots\cdots & & & 1 \end{bmatrix}$$

Such controllers are suitable for improving integrity when only subsystems require modification and in reducing interaction by diminishing the magnitudes of the off-diagonal elements of the plant $\underline{G}_p(s)$.

b) Scalar

$$\underline{G}_{c_i}(s) = k(s)\underline{I}$$

This controller multiplies the plant characteristic loci (each eigenvalue) by the scalar, $k(s)$ while leaving the characteristic directions unchanged.

c) Permutation

$$\underline{G}_{c_i}(s) = [e_1 \ldots e_{q-1} e_p e_{q+1} \ldots e_{p-1} e_q e_{p+1} \ldots e_m]$$

where $p > q$ and e_j is column j of $\underline{I}$. A controller of this form interchanges columns p and q of the plant matrix, $\underline{G}_p(s)$ which may be helpful in improving integrity.

d) Proportional plus Integral Action

$$\underline{G}_{c_i}(s) = \underline{K}_1 \underline{D}_1 + \underline{G}_p^{-1}(0)\underline{D}_2/s$$

The matrix $\underline{K}_1$ tends to render $\underline{G}_p(s)$ diagonal as $|s| \to \infty$, so tends to align the characteristic directions of $\underline{G}_p(s)$ with the standard basis vectors. Matrices $\underline{D}_1$ and $\underline{D}_2$, which are diagonal, can be used to adjust the weighting between zero and infinite frequencies in each column of $\underline{Q}(s) = \underline{G}_{c_i}(s)\underline{G}_p(s)$. Consequently the controller eliminates steady-state error by ensuring that at low frequencies, the moduli of all characteristic loci are large as $|s| \to 0$. High frequency interaction is also reduced by use of a sub-controller of this form.

The very nature of the iterative characteristic loci design procedure, as is the case for other multivariable frequency domain design techniques, may in fact limit its use. This is because to effectively utilize the technique requires the use of a digital computer with a visual display unit. Not to mention the vast number of man-hours required to develop the necessary software for implementation. Notwithstanding this limitation, to gain some appreciation of the actual procedure involved in designing the series of sub-controllers the various phases that are involved will be summarized:

a) <u>Stability phase</u> - This involves determination of the right-half plane zeros in the open loop characteristic polynomial. Closed loop stability is then assessed for a gain, k, applied to each loop. This is done by inspecting a display of the loci of $\underline{G}_p(j\omega)$ in the form of a Nyquist plot relative to the critical point $(-1/k, 0)$ for a finite number of frequencies.

b) <u>Integrity phase</u> - The same procedure as employed in the stability phase except that the encirclement theorem (stability check) is applied to the characteristic loci of the principal submatrices of $\underline{Q}(j\omega)$. Specifically when applied to the diagonal element $q_{ii}(j\omega)$ of $\underline{Q}(j\omega)$, the theorem establishes the stability margin when all loops except loop "i" are open. This analysis plus that of the stability phase provides an excellent indication of the stable operating regions for all possible combinations of loop gains k_i, $i = 1, 2, \ldots m$. Should integrity be poor, a controller factor $k_i(s)$ is synthesized and the stability phase is then repeated.

c) <u>Interaction phase</u> - Once the stability and integrity requirements have been satisfied the amount of interaction can be assessed from plots of $|q_i(j\omega)|$ versus ω and $\theta_i(j\omega)$ versus ω for $i = 1, 2, \ldots m$ where the set $\{q_i(j\omega)\}$ represents the characteristic loci of $\underline{Q}(j\omega)$ and $\theta_i(j\omega)$ is the minimum angle of misalignment between the standard basis vector e_i and the characteristic directions, $w_j(\overline{j\omega})$ of $\underline{Q}(j\omega)$ for all j. Interaction can be suppressed by insuring that either $|q_i(j\omega)| \gg 1$ or $\theta_i(j\omega) \approx 0$ for $i = 1, 2, \ldots m$. If the degree of interaction is not acceptable, compensation by means of a controller factor $\underline{\underline{G}}_{c_i}(s)$ is designed and the stability phase is repeated.

d) <u>Performance phase</u> - Finally when the major design considerations of stability, integrity, and interaction have been satisfied, compensation by applying single loop techniques to the diagonal elements of $\underline{Q}(s)$ can be undertaken. It is in this phase that the final loop gain values are tuned to give the controller factor $\underline{\underline{G}}_{c_i}(s) = \text{diag } \underline{\underline{G}}_{c_i}$.

As noted previously such a design procedure involves man-computer interaction by means of a visual display unit. Consequently, the experience of the control engineer will significantly influence the detailed steps and types of sub-controllers employed in achieving a final control system design. The reader interested in gaining a further appreciation of this technique should study the examples presented by Belletrutti (8) and Belletrutti and MacFarlane (7,10). Unfortunately because of the large amount of software development that is necessary to employ this, or for that matter any of the other multivariable frequency domain design techniques, the adoption of such design procedures is likely to be slow.

CONCLUSION

The improved control behaviour that can be achieved using combined feedforward-feedback control is well known so consequently the specification of such a control strategy is becoming commonplace in most organizations. However, this is not so for the multivariable frequency domain design techniques for developing control schemes. The non-interacting/decoupling design approach has found some application to industrial control problems but not the characteristic loci, or for that matter either of the Nyquist array techniques. This was the finding of Rijnsdorp and Seborg (3) in a survey prepared for the Engineering Foundation Conference on Chemical Process Control held in January, 1976. This lack of acceptance is in part due to the investment of time and resources required to implement the software to effectively utilize these design procedures. Furthermore, with the present lack of experimental evaluations of these techniques, such as the study of Kuon (22), the reluctance of designers to employ such techniques is understandable.

With the availability of process control computers has come the development of robust on-line adaptive techniques which involve estimation and control. Typical of these single variable techniques are the self-tuning regulator developed by Åström and co-workers (39) at the Lund Institute of Technology in Sweden and the self-tuning controller of Clarke and Gawthrop (40). Successful industrial applications of the self-tuning regulator approach have been reported in the mineral (41) and pulp and paper (42) industries. Availability of process control computers has also seen the development of multivariable control strategies based on steady state models. Using the data acquisition capability of the computer, static model calculations are made at frequent intervals (e.g. a few seconds) and the set points of several control loops adjusted. Also employed are on-line estimation procedures for periodic updating of model parameters.

REFERENCES

(1) Buckley, P.S., "Techniques of Process Control", John Wiley & Sons, Inc., New York (1964).

(2) Fisher, D.G. and Seborg, D.E., "Multivariable Computer Control, A Case Study", North-Holland Publishing Co., New York (1976).

(3) Rijnsdorp, J.E. and Seborg, D.E., "A Survey of Experimental Applications of Multivariable Control to Process Control Problems", AIChE Symposium Series, No. 159, 72 (to be published in 1977).

(4) Rosenbrock, H.H., "Design of Multivariable Control Systems Using the Inverse Nyquist Array", Proc. IEE, 116, 1929-1936 (1969).

(5) McMorran, P.D., "Extension of the Inverse Nyquist Method", *Electron. Lett.*, 6, 800-801 (1970).

(6) Rosenbrock, H.H., "Progress in the Design of Multivariable Control Systems", *Meas. and Contr.* 4, 9-11 (1971).

(7) Belletrutti, J.J. and MacFarlane, A.G.J., "Characteristic Loci Techniques in Multivariable-Control-System Design", *Proc. IEE*, 118, 1291-1298 (1971).

(8) Belletrutti, J.J., "Computer-aided Design and the Characteristic Locus Method", *Conference on Computer-Aided Control System Design*, IEE Conf. Pub. 96, 79-86 (1973).

(9) MacFarlane, A.G.J., "Frequency Response Methods in Multivariable Feedback System Design", *Conference on Computer-Aided Control System Design*, IEE Conf. Pub. 96, 71-78 (1973).

(10) MacFarlane, A.G.J. and Belletrutti, J.J., "The Characteristic Locus Design Method", *Automatica*, 9, 575-588 (1973).

(11) MacFarlane, A.G.J., "Return-difference and Return-ratio and their Use in the Analysis and Design of Multivariable Feedback Control System", *Proc. IEE*, 117, 2037-2049 (1970).

(12) Rosenbrock, H.H., *Recent Mathematical Developments in Control*, Ed. by D.J. Bell, Academic Press, London, 345 (1973).

(13) Rosenbrock, H.H., *Computer-Aided Control System Design*, Academic Press, New York (1974).

(14) Layton, J.M., "Commutative Controller: A Critical Survey", *Electron. Lett.*, 6, 362-363 (1970).

(15) MacFarlane, A.G.J., "Commutative Controller: A New Technique for the Design of Multivariable Control Systems", *Electron. Lett.*, 6, 121-123, 363-364 (1970).

(16) Owens, D.H., "Dyadic Approximation Method for Multivariable Control Systems Analysis with a Nuclear Reactor Application", *Proc. IEE*, 120, 801-809 (1973).

(17) Mayne, D.Q., "The Design of Linear Multivariable Systems", *Proc. 5th IFAC Congress*, Paris, 29.1 (1972).

(18) Mayne, D.Q., "The Design of Linear Multivariable Systems", *Automatica*, 10, 405-412 (1974).

(19) Mayne, D.Q., "The Effect of Feedback on Linear Multivariable Systems", *Automatica*, 10, 405-412 (1974).

(20) Owens, D.H., "Dyadic Modification to Sequential Technique for Multivariable Control Systems Design", *Electron. Lett.*, 10, 25-26 (1974).

(21) Owens, D.H., "Sequential Design of Linear Multivariable Systems Retaining Full Output Feedback", *Electron. Lett.*, 10, 79-80 (1974).

(22) Kuon, J.F., "Multivariable Frequency-Domain Design Techniques", Ph.D. Thesis, University of Alberta, Edmonton, Canada (1975).

(23) Denn, M.M. and Foss, A.S., "Perceptions of the State of Process Control", *AIChE J.*, 22, 1157-1158 (1976).

(24) Lupfer, D.E. and Parsons, J.R., "A Predictive Control System for Distillation Columns", *Chem. Eng. Progr.*, 58, (9), 37-42 (1962).

(25) MacMullan, E.C. and Shinskey, F.G., "Feedforward Analog Computer Control of a Superfractionator", *Cont. Eng.*, 11, (3), 69-74 (1964).

(26) Nisenfeld, A.E. and Miyasaki, R.K., "Applications of Feedforward Control to Distillation Columns", *Automatica*, 9, 319-327 (1973).

(27) Shinskey, F.G., *Process Control Systems*, McGraw-Hill, Inc., New York (1967).

(28) Wood, R.K. and Pacey, W.C., "Experimental Evaluation of Feedback, Feedforward and Combined Feedforward-Feedback Binary Distillation Column Control", *Can. J. Chem. Eng.*, 50, 376-384 (1972).

(29) Davis, T.P. and Smith, C.A., "An Evaluation of Different Control Schemes for Viscosity Control of a Fuel Oil Heating System", *ISA/77*, Anaheim, California (1977).

(30) Bristol, E.H., "On a New Measure of Interaction for Multivariable Process Control", *IEE Trans. Aut. Cont.*, AC-11, 133-134 (1966).

(31) Boksenbom, A.S. and Hood, R., "General Algebraic Method Applied to Control Analysis of Complex Engine Types", Report NCA-TR-980, National Advisory Committee for Aeronautics, Washington, D.C. (1949).

(32) Rosenbrock, H.H., "On the Design of Linear Multivariable Control Systems", *Proc. 3rd IFAC Congress*, London, 1-A1-A16 (1966).

(33) MacFarlane, A.G.J., "A Survey of Some Results in Linear Multivariable Feedback Theory", *Automatica*, 8, 455-492 (1972).

(34) Zalkind, C.S., "Practical Approach to Non-Interacting Control" (Part I,II), *Inst. Cont. Systems*, 40, No. 3, 89-93, No. 4, 111-116 (1967).

(35) Luyben, W.L., "Distillation Decoupling", *AIChE J.*, 16, 198-203 (1970).

(36) Wood, R.K. and Berry, M.W., "Terminal Composition Control of a Binary Distillation Column", *Chem. Eng. Sci.*, 28, 1707-1717 (1973).

(37) MacFarlane, A.G.J., "Relationships between Recent Developments in Linear Control Theory and Classical Design Techniques (Parts 1-5)", *Meas. and Contr.*, 8, 179-187, 219-223, 278-284, 319-323, 371-375 (1975).

(38) Schwanke, C.O., Edgar, T.F. and Hougen, J.O., "Development of Multivariable Control Strategies for Distillation Columns", *ISA/77*, Anaheim, California (1977).

(39) Åström, K.J. and Wittenmark, B., "On Self Tuning Regulators", *Automatica*, 9, 185-199 (1973).

(40) Clarke, D.W. and Gawthrop, P.J., "Self-tuning Controller", *Proc. IEE*, 122, 929-934 (1975).

(41) Borrison, U and Syding, R., "Self-tuning control of an ore crusher", *Automatica*, 12, 1-7 (1976).

(42) Cegrell, T. and Hedquist, T., "Successful Adaptive Control of Paper Machines", *Automatica*, 11, 53-59 (1975).

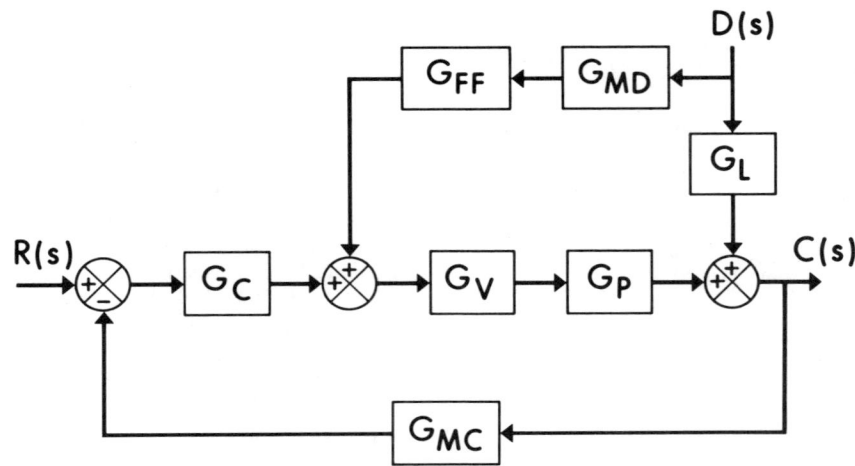

Figure 1. Block Diagram of Combined Feedforward-Feedback Control System

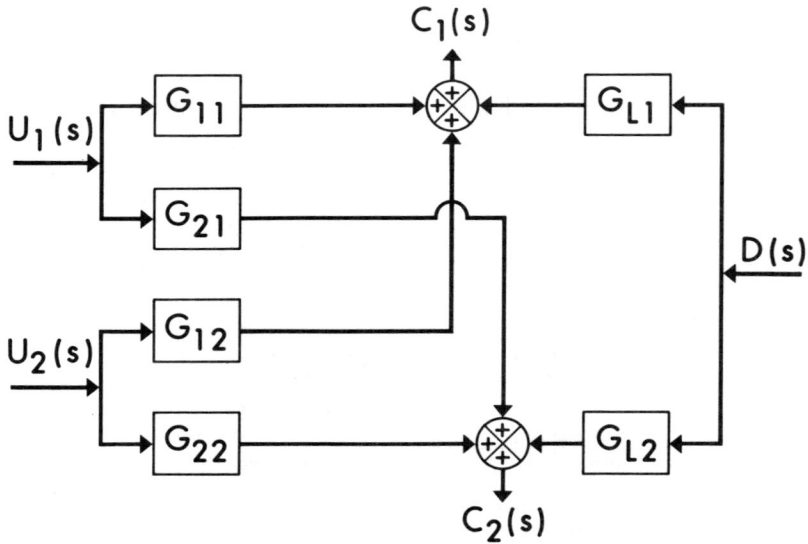

Figure 2. Block Diagram of Multivariable System

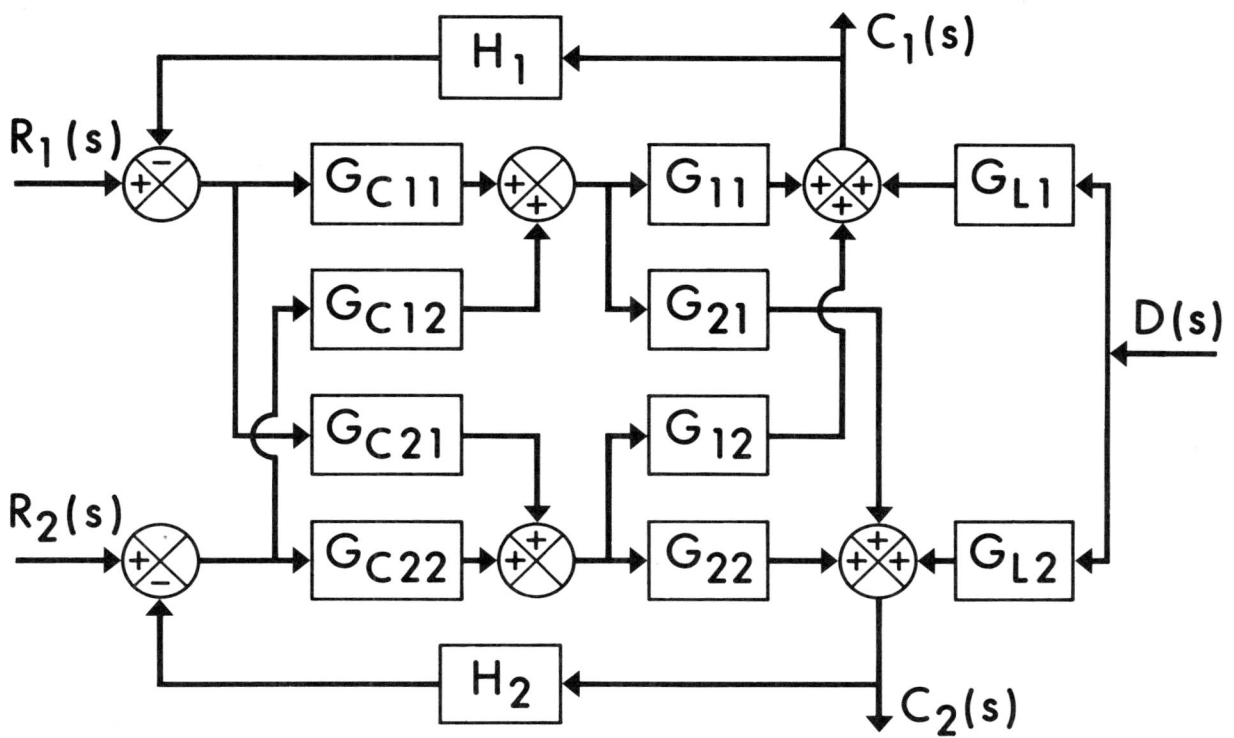

Figure 3. Block Diagram of Non-interacting Control System

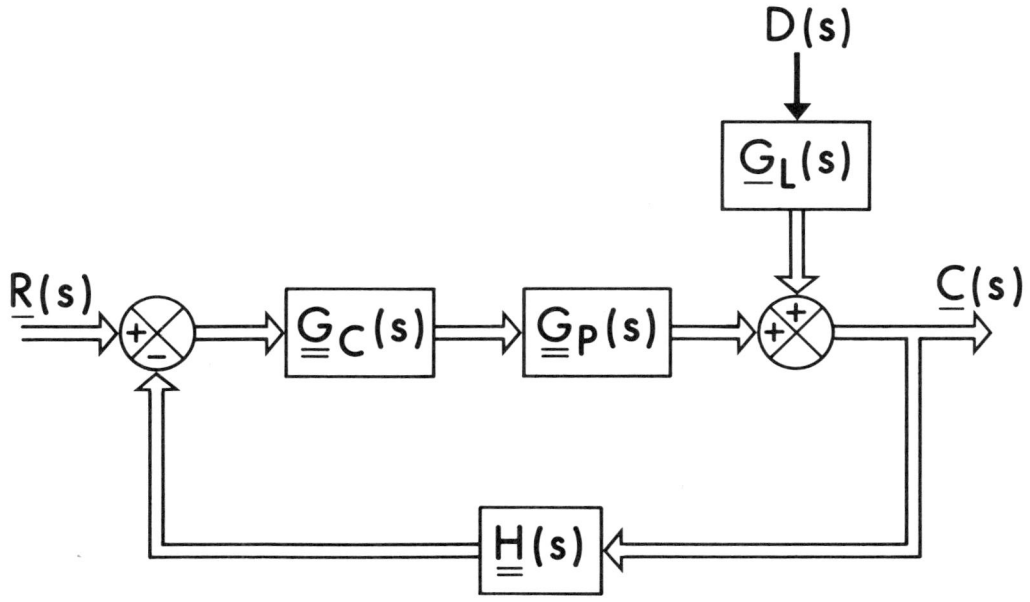

Figure 4. General Block Diagram for Multivariable Feedback System

©ISA, 1977
ISBN 87664-363-2

A COMPARISON OF CONTROLLER ALGORITHMS AS APPLIED TO A STIRRED TANK REACTOR

Benjamin T. Condon
Monsanto Chemical Company
Lulling, Louisiana 70070

Carlos A. Smith
University of South Florida
Chemical Engineering
Tampa, Florida 33620

INTRODUCTION

The purpose of this paper is to present a comparison of three different control algorithms as applied to a simulated chemical reactor. These three algorithms are a continuous proportional-integral-derivative (PID) controller, a discrete proportional-integral-derivative controller and Dahlin's digital controller. The discussion starts with the description of the simulated system and continues with the basic PID continuous controller, the discrete PID controller and the development of the Dahlin's controller. Finally, each control algorithm is evaluated according to how well it controls the outlet concentration of the reactor.

SYSTEM DESCRIPTION

The simulated system consisted of a continuous stirred tank reactor, Fig. 1, where an exothermic reaction of the type A→B takes place. The controlled variable of interest was the outlet concentration of component A. The control was accomplished by sensing the outlet concentration of component A and adjusting the water flow rate, manipulated variable, through the cooling jacket. The adjusting was done according to the particular control algorithm being used. For the specifications of the reactor and related control hardware the reader is referred to Appendix A.

PID CONTINUOUS CONTROLLER

The controller selected to compare the two discrete controllers which was the continuous pneumatic PID controller. The equation used to represent the action of this controller was

$$m(t) = K_c(e(t) + \frac{1}{\tau_I}\int e(t)dt + \tau_D \frac{de(t)}{dt}) + m_0 \quad (1)$$

PID DISCRETE CONTROLLER

This controller is derived directly from equation (1). This is done by representing the integral with a rectangular approximation and the derivative with a finite backwards difference.

$$m_n = K_c(e_n t + \frac{T}{\tau_I}\sum_{i=1}^{N} e_i + \frac{\tau_D}{T}(e_n - e_{n-1})) + m_0 \quad (2)$$

By inspection it can be seen that equation (2) offers no protection from "reset windup" (integration saturation) and also needs to be initialized. To overcome these disadvantages, equation (2) can be rewritten for m_{n-1}, the controller's output for the previous sampling time.

$$m_{n-1} = K_c(e_{n-1} + \frac{T}{\tau_I}\sum_{i=1}^{n-1} e_i + \frac{\tau_D}{T}(e_{n-1} - e_{n-2})) + m_0 \quad (3)$$

By subtracting equation (3) from (2) a new equation develops:

$$\Delta m_n = m_n - m_{n-1} = K_c((e_n - e_{n-1}) + \frac{T}{\tau_I}e_n + \frac{\tau_D}{T}(e_n - 2e_{n-1} + e_{n-2})) \quad (4)$$

Equation (4) is called the "velocity form" of the PID discrete controller. It is probably one of the most commonly used algorithms in computer control loops. It provides "anti-reset windup protection" and a bumpless transfer.

The computer control system schematic used in this work is shown in Fig. 2. The block diagram is shown in Fig. 3.

DAHLIN'S CONTROLLER

One of the most advanced control algorithms now available for use in computer control installations is the one developed by Dahlin[1,2]. This computer algorithm is the second discrete controller selected for analysis. The form of this algorithm is:

$$m_n = \frac{1}{\gamma_4}\left[\frac{1}{K_S}(\gamma_2 e_n + K_1 e_{n-1} - K_2 e_{n-2}) - K_3 m_{n-1} + K_4 m_{n-2} + K_5 m_{n-N-1} + K_6 m_{n-N-2} + K_7 m_{n-N-3}\right] \quad (5)$$

For the development of this algorithm and for the definition of the terms the reader is referred to Appendix B.

COMPARISON AND DISCUSSION

Each control algorithm presented, equations (1), (4) and (5), was used to control the chemical reactor. The closed loop system was implemented for digital simulation in an IBM 360 computer using the Continuous System Modeling Program[3], CSMP.

After optimum tuning, each control algorithm system was subjected to a step change in set point. Each system's response is recorded in Fig. 4. It is apparent from this figure that, for this case, the PID continuous controller performed "better" than the discrete controllers.

To further investigate the relative merits of the various control algorithms, different amounts of dead time were added to the system. This can be done by physically moving the concentration sensor away from the reactor along the outlet pipe. Each control algorithm was again optimally tuned for every run. The integral of the Absolute Value of the Error (IAE) was the criterion for performance. Figures (5) and (6) show the process output as a function of time for runs with 100 and 200 seconds of added dead time respectively. From these figures it becomes apparent that as the dead time of the process is increased, the relative merits of the discrete controllers also increase.

This observation is further substantiated by Fig. 7 where the integral of the absolute value of the error (IAE) is plotted, for each run, versus the total dead time. This IAE serves as the criterion to compare all three different responses. This figure clearly indicates that for this system as dead time increases the performance of the discrete algorithms surpass that of the continuous controller. Also, the performance of Dahlin's algorithm is superior to that of the discrete PID controller.

It can thus be concluded, that for this system, the use of Dahlin's algorithm may yield significant process control improvement over other control techniques if the dead time is a significant factor. The point at which Dahlin's algorithm becomes superior is of course peculiar to the specific applications.

NOMENCLATURE

$e(t)$	controller's error = set point - measured variable, psig
e_n	present error for discrete controller
K_c	controller's gain, psig/psig
$m(t)$	continuous controller's output, psig
m_n	present output from discrete controller
m_{n-1}	previous output from discrete controller
m_0	controller's bias value, psig
T	sampling time, sec
τ_I	reset time, min/repeat
τ_D	rate time, minutes

APPENDIX A: REACTOR MODELING, SPECIFICATIONS AND SIMULATION

This appendix presents the description of the stirred tank chemical reactor used in this work. Specifically, the modeling, specification and computer simulation of the reactor are presented. The response of this reactor is quite nonlinear arising mainly from the Arrhenius term in the kinetic expression.

The reaction being controlled is irreversible exothermic reaction of the type:

$$A \xrightarrow{k} B$$

with first oreer kinetics taking place. The reaction rate is given by:

$$r_A = k_0 \exp[E/R(T + 460.)]C_A \quad (A.1)$$

where:

r_A = reaction rate, moles of formed ft^3-hr.

k_0 = frequency factor, 7.08×10^{10} hr^{-1}

E = activation energy, 30000 Btu/mole

R = gas constant, 1.987 Btu/mole-$^\circ K$

T = temperature, $^\circ F$

C_A = concentration of A in the reactor, moles/ft^3

The mass balance on the reactor contents, assuming uniform mixing and constant density, is:

$$\rho F_0 - \rho F = \rho A \frac{dh}{dt} \quad (A.2)$$

where:

- F_0 = Inlet flow, 2000 ft^3/hr
- F = outlet flow, ft^3/hr
- A = reactor's inside cross sectional area, 5.0 ft^2
- h = level in reactor, ft.
- ρ = liquid density, assumed constant, 50 lb$_m$/ft^3

The mass balance on component A is:

$$F_0 C_{A0} - F C_A - A h\, r_A = A \frac{d(h C_A)}{dt} \quad (A.3)$$

where:

- C_{A0} = inlet concentration of A, 0.4 moles/ft^3

Since the reactor is assumed to be well mixed, the reactor's content concentration equals the outlet concentration.

The heat balance on the reactor's contents is:

$$F_0 \rho C_p (T_0 - T_R) - \Delta H_r A h r_A - h_i A_i (T - T_m)$$
$$- F \rho C_p (T - T_R)$$
$$= \rho A C_p \frac{d[h(T - T_R)]}{dt} \quad (A.4)$$

where:

- C_p = liquid heat capacity, assumed constant, 0.75 Btu/lb$_m$ °F
- T_0 = reactant's inlet temperature, 250 °F
- T_R = reference temperature, 32.0 °F
- ΔH_r = heat of reaction, -1500 Btu/mole
- h_i = inside film heat transfer coefficient, assumed constant, 160 Btu/hr-ft^2-°F
- A_i = inside heat transfer area, assumed constant, 250 ft^2
- T_m = metal wall temperature, °F
- T = reactor's content temperature, °F

The heat balance on the cooling water jacket is:

$$F_J \rho_J C_{pJ}(T_{J_0} - T_J) + h_0 A_0 (T_m - T_J)$$
$$= V_J \rho_J C_{pJ} \frac{d(T_J)}{dT} \quad (A.6)$$

where:

- F_J = volumetric flow rate of water through the cooling jacket, ft^3/hr
- ρ_J = water density, assumed constant, 62.4 lb$_m$-°F
- C_{pJ} = heat capacity of water, 1.0 Btu/lb$_m$-°F
- T_{J_0} = inlet water temperature, 70 °F
- V_J = jacket volume, 15.0 ft^3

The flow rate out of the reactor tank is calculated according to the following equation for flow through a valve:

$$F = 8.02\, C_v\, v_p \sqrt{\frac{\Delta p}{G}} \quad (A.7)$$

where:

- C_v = valve coefficient, 239.4
- v_p = valve position
- Δp = pressure drop over valve, psig assuming that the reactor is open to the atmosphere and that the valve discharges to the atmosphere, i.e.:

$$\Delta p = \frac{\rho g h}{g_c} = \rho h / 144 \text{ psia}$$

- G = specific gravity of reactor's contents, 0.8

The cooling water flow rate (F_J) is calculated using the same equation:

$$F_J = 8.02\, C_{vJ}\, v_p J \sqrt{\frac{\Delta p J}{G_J}} \quad (A.8)$$

where:

- C_{vJ} = water jacket valve coefficient, 34.8
- v_{pJ} = valve position of water jacket valve
- ΔpJ = pressure drop across the water jacket valve, 30 psi
- G_J = water specific gravity, 1.0

In addition to the above equations that fully describe the reactor, some other descriptive equations are necessary to complete the control loop.

In order to measure and transmit the outlet reactor's concentration to the concentration controller, a concentration transmitter is used. This transmitter is calibrated for a range of 0 - 1.5 moles/ft^3. Since this is a pneumatic transmitter, its output range is 3 - 15 psig. This calibration allows us to obtain the transmitter gain, K_{CT}:

$$K_{CT} = \frac{(15-3) \text{ psig}}{(1.5-0) \text{ moles/ft}^3} = 8.0 \text{ psig/moles/ft}^3$$

The descriptive equation for this transmitter is:

$$C_{AT} = 3.0 + K_{CT}(C_A) \quad (A.9)$$

where:

C_{AT} = concentration of component A as transmitted by the transmitter, psig

The level transmitter used to measure the liquid in the reactor has a calibration of 0-20 ft. for an output of 3-15 psig. Therefore, this transmitter gain, K_{LT}, is:

$$K_{LT} = \frac{(15-3) \text{ psig}}{(20-0) \text{ ft}} = 0.6 \text{ psig/ft}$$

The descriptive equation is:

$$h_T = 3.0 + K_{LT}(h) \quad (A.10)$$

where:

h_T = transmitted level, psig

The cooling water valve actuator can be described as:

$$F_J vp = K_v(m_c) - 0.25 \quad (A.11)$$

where:

$F_J vp$ = outlet from actuator to the valve, valve position

m_c = input to actuator from concentration controller, psig

K_v = actuator's gain, valve position/psig

The liquid level valve actuator is also described as follows:

$$F_L vp = K_v(m_h) - 0.25 \quad (A.12)$$

where:

$F_L vp$ = output from the actuator to the valve, valve position

m_h = input to actuator from level controller, psig

APPENDIX B

The derivation of Dahlin's algorithm begins from the analysis of the block diagram in Fig. B.1. The procedure used is analogous to that used to analyze the block diagram for a continuous system except that z-transforms are used.

The analysis begins by relating $C(z)$ and $R(z)$ through the closed loop pulse transfer function, i.e.:

$$\frac{C(z)}{R(z)} = \frac{D(z) HG(z)}{1 + D(z) HG(z)}$$

where:

$C(z)$ = z-transformation of the sampled function of the controlled variable.

$R(z)$ = z-transformation of the sampled function of the controller's set point.

$D(z)$ = controller's pulse transfer function

$HG(z)$ = pulse transfer function of the zero-order hold and process.

Solving the above transfer function for $D(z)$ gives

$$D(z) = \frac{M(z)}{E(z)} = \frac{1}{HG(z)} \cdot \frac{C(z)/R(z)}{1 - C(z)/R(z)} \quad (B.1)$$

where:

$M(z)$ = z-transformation of discrete controller's output function

$E(z)$ = z-transformation of sampled error function.

By inspection it becomes apparent from equation (B.1) that if $C(z)$, $R(z)$ and $HG(z)$ could somehow be specified, it would be possible to design the control algorithm for the specified set of conditions $C(z)$, $R(z)$ and $HG(z)$.

In the following sections $C(z)$, $R(z)$ and $HG(z)$ are specified and the resulting control algorithm, $D(z)$, developed. Equation (B.1) is referred to as the "synthesis equation" for digital controller's design.

A. Set Point Function, $R(z)$

$R(z)$ is the z-transformation of the sampled set point function. This analysis is limited to set point changes, therefore:

$$R(z) = z\left[\frac{1}{s}\right] = \frac{1}{1-z^{-1}} \quad (B.2)$$

This is a realistic specification since many set point changes occur in this manner.

B. Process Output, $C(z)$

The basis of Dahlin's digital controller is that the closed loop response, to a step change in set point, should be a first order lag plus dead time[1]. That is, the closed loop response, in the Laplace domain, is:

14

$$C(s) = \frac{e^{-\theta_L s}}{\lambda s + 1} \cdot \frac{1}{s}$$

where:

θ_L = dead time of the closed loop system

λ = time constant of the closed loop system; this is Dahlin's tuning parameter.

Obtaining the z-transform of $C(s)$:

$$C(z) = z\left[\frac{e^{-\theta_L s}}{\lambda s + 1} \cdot \frac{1}{s}\right]$$

If θ_L is not an integer number of sampling time, T, then the modified z-transforms are used[4]. Let:

$$\theta_L = NT + \theta_L'$$

where:

N = closest integer number of sampling times in the loop's dead time.

then:

$$C(z) = z\left[\frac{e^{-NTs} e^{-\theta_L' s}}{\lambda s + 1} \cdot \frac{1}{s}\right]$$

$$C(z) = z^{-N} z\left[\frac{e^{-\theta_L' s}}{\lambda s + 1} \cdot \frac{1}{s}\right]$$

or:

$$C(z) = z^{-N} z_m\left[\frac{1}{\lambda s + 1} \cdot \frac{1}{s}\right]$$

which yields:

$$C(z) = z^{-N-1}\left[\frac{(e^{-mT/\lambda} - e^{-T/\lambda})z^{-1} + (1 - e^{-mT/\lambda})}{(1 - z^{-1})(1 - e^{-T/\lambda} z^{-1})}\right]$$

where:

$$m = 1 - \frac{\theta_L'}{T}$$

Since $R(z)$ has been specified, then $\frac{C(z)}{R(z)}$ can be evaluated:

$$\frac{C(z)}{R(z)} = Z^{-N-1}\left[\frac{\gamma_1 z^{-1} + \gamma_2}{(1 - e^{-T/\lambda} z^{-1})}\right]$$

(B.3)

where:

$$\gamma_1 = e^{-mT/\lambda} - e^{-T/\lambda}$$

$$\gamma_2 = 1 - e^{-mT/\lambda}$$

C. Process Transfer Function, G(s)

In order to use the synthesis equation (B-1) it is necessary to determine HG(z). To obtain this it is necessary to obtain G(s), the open loop transfer function. Possibly the best method to obtain G(s) is by the process reaction method. A process reaction curve is shown in Fig. B-2. One easily obtained from an actual process by manually altering the manipulated variable during steady state operation and recording the resulting response as detected by the transmitter. This process curve is usually approximated as a first order lag plus dead time (5,6). The form of this is:

$$G(s) = \frac{K_s e^{-\theta_s s}}{\tau_s s + 1}$$

where:

K_s = process' gain, psig/psig

τ_s = process' time constant, sec.

θ_s = process' dead time, sec.

Knowing the process transfer function allows us to evaluate the pulse transfer function HG(z)

$$HG(z) = z[H(s)G(s)]$$

where:

H(s) = zero-order hold transfer function

$$H(s) = \frac{1 - e^{-sT}}{s}$$

then:

$$HG(s) = z\left[\frac{1 - e^{-sT}}{s} \cdot \frac{K_s e^{-\theta_s s}}{\tau s + 1}\right]$$

Let: $\theta_s = NT + \theta_s'$

Then using modified z-transform it yields:

$$HG(z) = K_s(1 - z^{-1})z^{-N-1}\left[\frac{\gamma_3 z^{-1} + \gamma_4}{(1 - z^{-1})(1 - e^{-T/\tau_s} z^{-1})}\right]$$

(B.4)

where:

$$m = 1 - \frac{\theta_s'}{T}$$

$$\gamma_3 = e^{-mT/\tau_s} - e^{-T/\tau_s}$$

$$\gamma_4 = 1 - e^{-mT/\tau_s}$$

It must be recognized that θ_L should be specified as close as possible to θ_s. If θ_L is specified less than θ_s then the resulting controller is physically unrealizable. If θ_L is specified greater than θ_s then dead time is added to the system.

D. Control Algorithm, D(z)

Knowing HG(z), R(z) and C(z) it is now possible to proceed with the development of D(z). Substituting (B.3) and B.4) into the synthesis equation (B.1).

$$D(z) = \frac{(1-z^{-1})(1-e^{-T/\tau_s}z^{-1})}{K_s z^{-N-1}(1-z^{-1})(\gamma_3 z^{-1}+\gamma_4)}$$

$$\frac{\frac{(z^{-N-1})(\gamma_1 z^{-1}+\gamma_2)}{(1-e^{-T/\tau_s}z^{-1})}}{1 - \frac{(z^{-N-1})(\gamma_1 z^{-1}+\gamma_2)}{(-e^{-T/\lambda}z^{-1})}}$$

which can be simplified to:

$$D(z) = \frac{(\frac{1}{K_s})(\gamma_2 + K_1 z^{-1} - K_2 z^{-2})}{(\gamma_4 + K_3 z^{-1} - K_4 z^{-2} - K_5 z^{-N-1} - K_6 z^{-N-2} - K_7 z^{-N-3})}$$

(B.5)

where:

$K_1 = \gamma_1 - \gamma_2 e^{-T/\tau_s}$

$K_2 = \gamma_1 e^{-T/\tau_s}$

$K_3 = \gamma_3 - \gamma_4 e^{-T/\lambda}$

$K_4 = \gamma_3 e^{-T/\lambda}$

$K_5 = \gamma_2 \gamma_4$

$K_6 = \gamma_3 \gamma_2 + \gamma_4 \gamma_1$

$K_7 = \gamma_3 \gamma_1$

but since $D(z) = \frac{M(z)}{E(z)}$ the above equation can be be rewritten as:

$M(z)(\gamma_4 + K_3 z^{-1} - K_4 z^{-2} - K_5 z^{-N-1} - K_6 z^{-N-2} - K_7 z^{-N-3})$

$= E(z)(\frac{1}{K_s})(\gamma_2 + K_1 z^{-1} - K_2 z^{-2})$

from this the following difference equation is obtained:

$\gamma_4 m_n + K_3 m_{n-1} - K_4 m_{n-2} - K_5 m_{n-N-1} - K_6 m_{n-N-2} - K_7 m_{n-N-3}$

$= (\frac{1}{K_s})(\gamma_2 e_n + K_1 e_{n-1} - K_2 e_{n-2})$

solving for m_n, the controller's output:

$m_n = (\frac{1}{\gamma_4})[(\frac{1}{K_s})(\gamma_2 e_n + K_1 e_{n-1} - K_2 e_{n-2}) - K_3 m_{n-1} + K_4 m_{n-2}$

$+ K_5 m_{n-N-1} + K_6 m_{n-N-2} + K_7 m_{n-N-3}]$

which is equation (5).

REFERENCES

1. Dahlin, E. B., "Design and Tuning Digital Controllers, Part I", <u>Instrument and Control Systems</u>, Vol. 41, No. 6, 1968.

2. Dahlin, E. B., M. G. Horner, W. A. Wickstran and R. L. Ziemer, "Designing and Tuning Digital Controllers, Part II", <u>Instrument and Control Systems</u>, Vol. 41, No. 6, 1968.

3. System 360 Continuous System Modeling Program Users Manual, Program Number 36DA-CX-16X. (IBM Document No. GH20-0367-4) January, 1972.

4. Higham, J. D., "Single Term Control of First and Second Order Processes with Dead Time", <u>Control Engineering</u>.

5. Smith, C. L., "Digital Computer Process Control", Intex Educational Publishers, Scranton, PA., 1972.

6. Condon, B. T., "A Study of Digital Controllers Applied to a Chemical Reactor", Master's Thesis, University of South Florida, August, 1975.

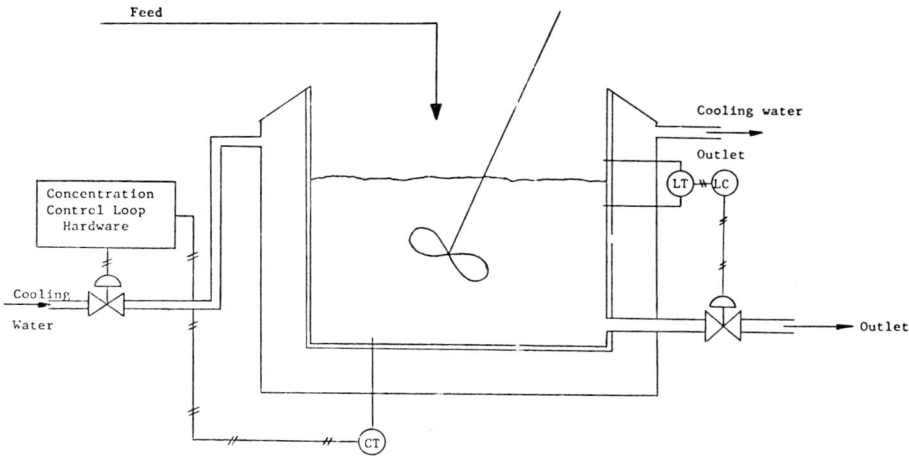

Figure 1   Stirred tank reactor with control loops.

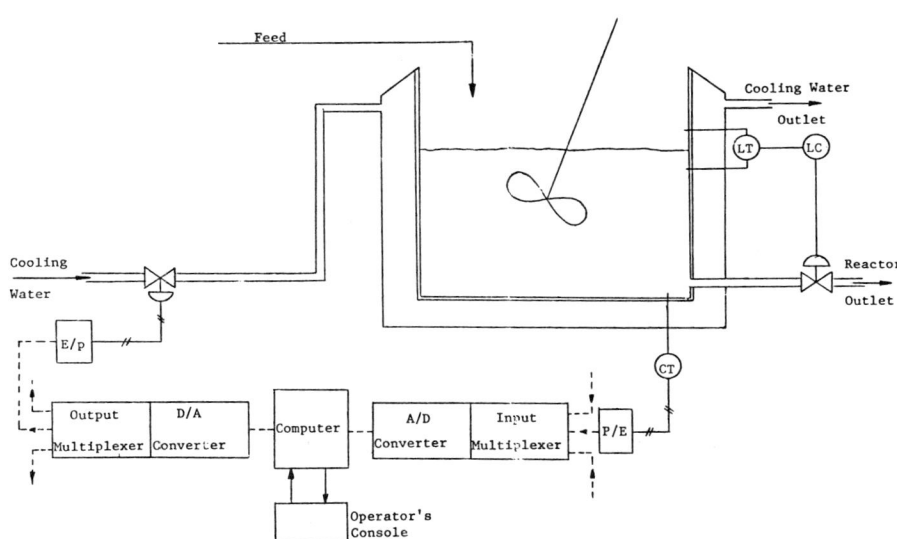

Figure 2   Direct Digital Control (DDC) System.

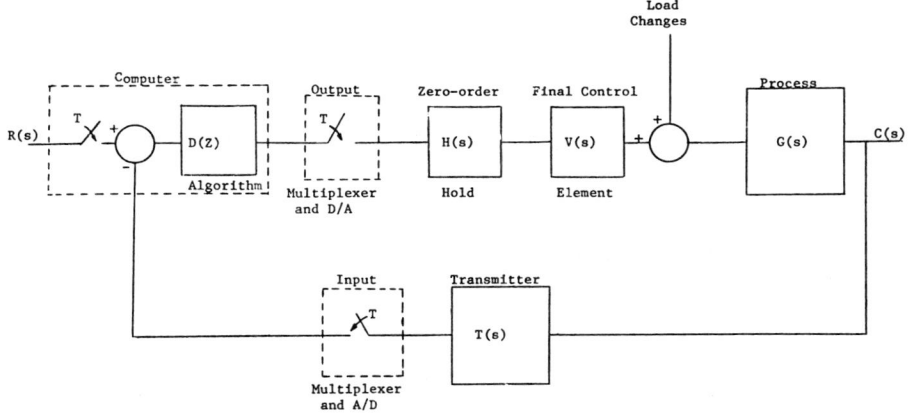

Figure 3   Block Diagram for a Discrete Control System.

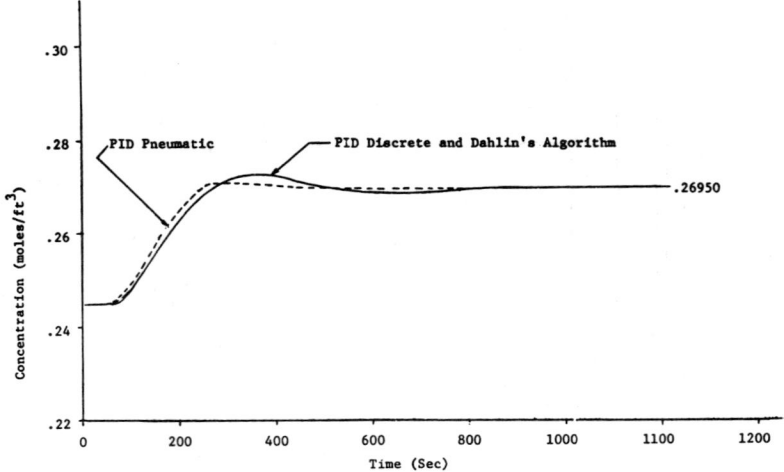

Figure 4   Reactor's concentration response for a 10% increase in controller's set point.

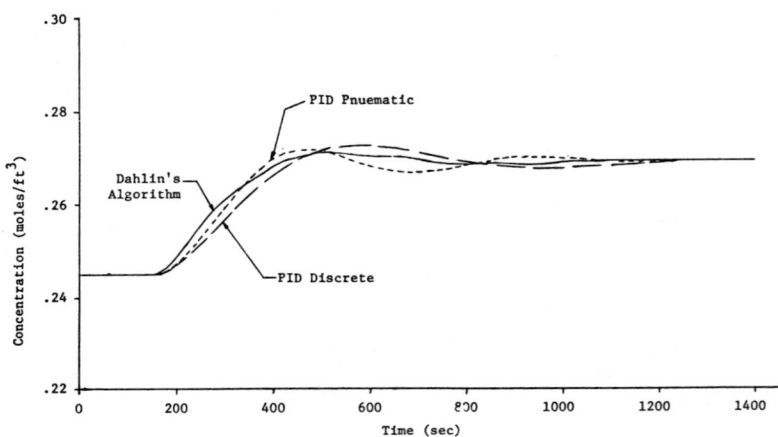

Figure 5   Reactor's Concentration Response to a 10% increase in controller's set point. 100 seconds added dead time.

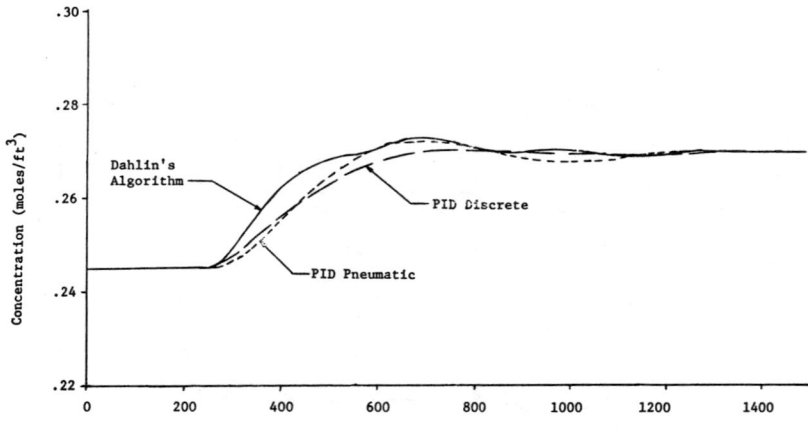

Figure 6   Reactor's Concentration Response to a 10% Increase in Controller's Set Point. 200 seconds added dead time.

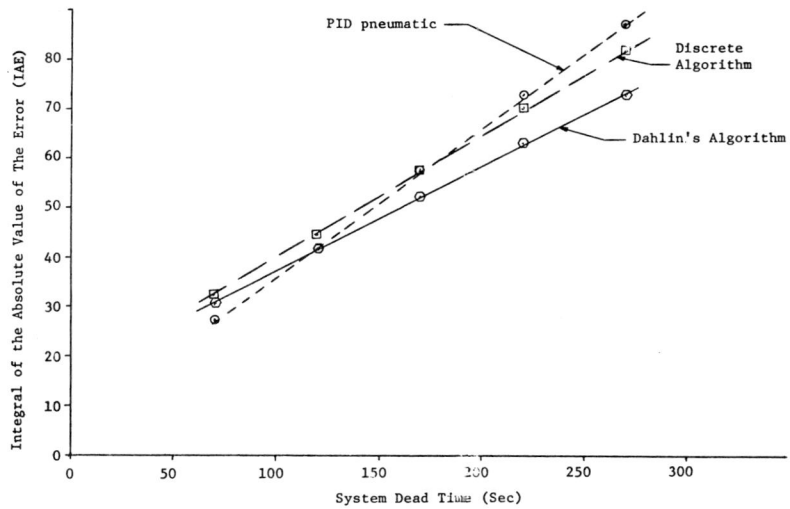

Figure 7 Comparison of three algorithms as Dead Time is Changed

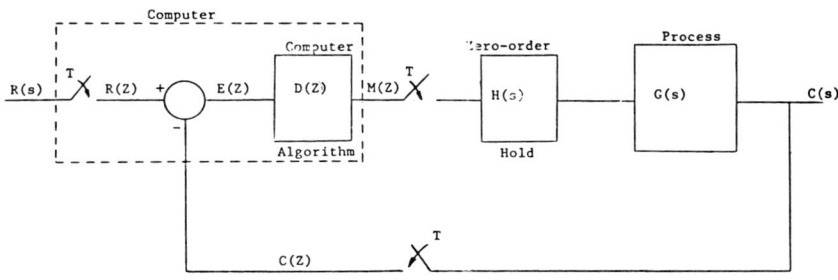

Figure B-1 Reduced Block Diagram of the Direct Digital Control System.

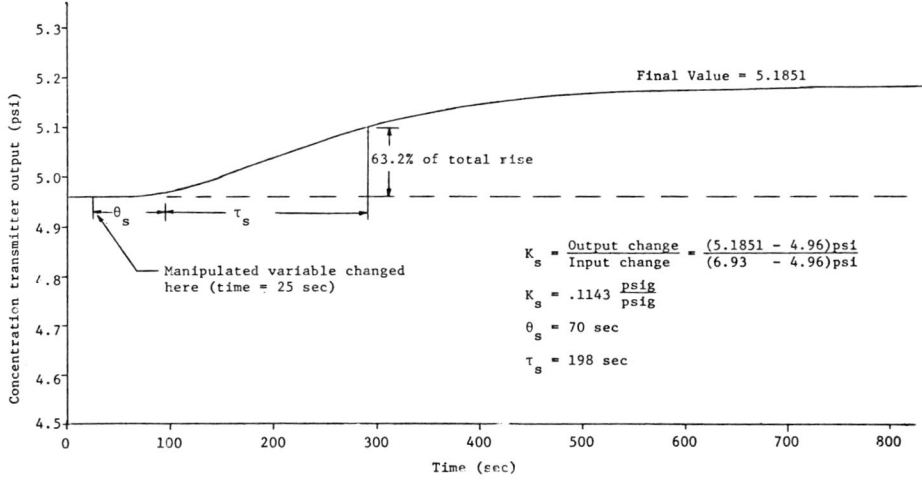

Figure B-2 Process reaction curve for reactor (valve actuator output changed from 4.96 psia to 6.93 psia)

A SENSITIVITY ANALYSIS ON DAHLIN'S

CONTROL ALGORITHM

Benjamin T. Condon
Process Engineer
Monsanto Company
Luling, La. 70070

Carlos A. Smith
Chemical Engineering
University of South Florida
Tampa, Florida 33620

INTRODUCTION

The purpose of this paper is to present a sensitivity analysis on Dahlin's control algorithm as applied to a simulated chemical reactor. The discussion starts with a description of the simulated system and continues with the development of the controller. Finally, the results of the sensitivity analysis are presented and evaluated.

SYSTEM DESCRIPTION

The simulated system consisted of a continuous stirred tank reactor, Figure 1, where an exothermic reaction of the type A→B takes place. The controlled variable of interest was the outlet concentration of component A. The control was accomplished by adjusting the water flow rate, manipulated variable, through the cooling jacket. For the specifications of the reactor and related control hardware the reader is referred to Appendix A.

CONTROL ALGORITHM

The derivation of the control algorithm starts from the analysis of the process's block diagram, Figure 2. The procedure used is analogous to that used to analyze the block diagram for a continuous system except that z-transforms are used.

The analysis begins by relating $C(z)$ and $R(z)$ through the closed loop pulse transfer function, i.e.:

$$\frac{C(z)}{R(z)} = \frac{D(z) HG(z)}{1 + D(z) HG(z)}$$

where:

$C(z)$ = z-transformation of the sampled function of the controlled variable

$R(z)$ = z-transformation of the sampled function of the controller's set point

$D(z)$ = controller's pulse transfer function

$HG(z)$ = pulse transfer function of the zero-order hold and process.

Solving the above transfer function for $D(z)$ gives

$$D(z) = \frac{M(z)}{E(z)} = \frac{1}{HG(z)} \cdot \frac{C(z)/R(z)}{1 - C(z)/R(z)} \quad (1)$$

where:

$M(z)$ = z-transformation of discrete controller's output function

$E(z)$ = z-transformation of sampled error function

By inspection it becomes apparent from equation (1) that if $C(z)$, $R(z)$ and $HG(z)$ could somehow be specified, it would be possible to design the control algorithm for the specified set of conditions $C(z)$, $R(z)$ and $HG(z)$. In the following paragraphs $C(z)$, $R(z)$ and $HG(z)$ are specified and the resulting control algorithm, $D(z)$, developed. Equation (1) is referred to as the "synthesis equation" for digital controller's design.

$R(z)$ is the z-transformation of the sampled set point function. This analysis is limited to set point changes, therefore:

$$R(z) = z\left[\frac{1}{s}\right] = \frac{1}{1-z^{-1}} \quad (2)$$

This is a realistic specification since many set point changes occur in this manner.

The basis of Dahlin's digital controller is that the closed loop response to a step change in set point should be a first order lag plus dead time[1]. That is, the closed loop response, in the Laplace domain, is:

$$C(s) = \frac{e^{-\theta_L s}}{\lambda s + 1} \cdot \frac{1}{s}$$

where:

θ_L = dead time of the closed loop system

λ = time constant of the closed loop system.

In his original article, Dahlin assumed θ_L to be an integer number of sampling time, T. In this paper, we do not assume this and, therefore, use the modified z-transform in the development of the controller. Obtaining the z-transform of C(s):

$$C(z) = z\left[\frac{e^{-\theta_L s}}{\lambda s + 1} \cdot \frac{1}{s}\right]$$

Let:

$$\theta_L = NT + \theta_L'$$

where:

  N = closest integer number of sampling times in the loop's dead time.

and:

$$0 < \theta_L' < T$$

then:

$$C(z) = z\left[\frac{e^{-NTs} e^{-\theta_L' s}}{\lambda s + 1} \cdot \frac{1}{s}\right]$$

$$C(z) = z^{-N} z\left[\frac{e^{-\theta_L' s}}{\lambda s+1} \cdot \frac{1}{s}\right]$$

$$C(z) = z^{-N} z_m\left[\frac{1}{\lambda s+1} \cdot \frac{1}{s}\right]$$

which yields:

$$C(z) = z^{-N-1} \frac{(e^{-mT/\lambda} - e^{-T/\lambda})z^{-1} + (1-e^{-mT/\lambda})}{(1-z^{-1})(1-e^{-T/\lambda}z^{-1})}$$

where:

$$m = 1 - \frac{\theta_L'}{T}$$

Since R(z) has been specified, the $\frac{C(z)}{R(z)}$ can be evaluated:

$$\frac{C(z)}{R(z)} = z^{-N-1} \frac{\gamma_1 z^{-1} + \gamma_2}{(1-e^{-T/\lambda}z^{-1})} \quad (3)$$

where:

$$\gamma_1 = e^{-mT/\lambda} - e^{-T/\lambda}$$

$$\gamma_2 = 1 - e^{-mT/\lambda}$$

In order to use the "synthesis equation" it is necessary to determine HG(z). To obtain this it is necessary to obtain G(s), the open loop transfer function. Possibly the best method to obtain G(s) is by the process reaction method. A process reaction curve is shown in Figure 3. One is easily obtained from an actual process by manually altering the manipulated variable during steady state operation and recording the resulting response as detected by the transmitter. This process curve is usually approximated as a first order lag plus dead time[2]. The form of this is:

$$G(s) = \frac{K_s e^{-\theta_s s}}{\tau_s s + 1}$$

where:

  K_s = process's gain, psig/psig

  τ_s = process's time constant, sec.

  θ_s = process's dead time, sec.

Knowing the process transfer function allows us to evaluate the pulse transfer function HG(z)

$$HG(z) = z\left[H(s)G(s)\right]$$

where:

  H(s) = zero-order hold transfer function

$$H(s) = \frac{1-e^{-sT}}{s}$$

then:

$$HG(s) = z\left[\frac{1-e^{-sT}}{s} \cdot \frac{K_s e^{-\theta_s s}}{\tau s+1}\right]$$

Let: $\theta_s = NT + \theta_s'$

Then using modified z-transform it yields:

$$HG(z) = K_s(1-z^{-1})z^{-N-1} \frac{\gamma_3 z^{-1} + \gamma_4}{(1-z^{-1})(1-e^{-T/\tau_s}z^{-1})}$$

(4)

where:

$$m = 1 - \frac{\theta_s'}{T}$$

$$\gamma_3 = e^{-mT/\tau_s} - e^{-T/\tau_s}$$

$$\gamma_4 = 1 - e^{-mT/\tau_s}$$

It must be recognized that θ_L should be specified as close as possible to θ_s. If θ_L is specified less than θ_s then the resulting controller is physically unrealizable. If θ_L is specified greater than θ_s then dead time is added to the system. Therefore, the most reasonable specification is to set $\theta_L = \theta_s$.

Knowing $HG(z)$, $R(z)$ and $C(z)$ it is now possible to proceed with the development of $D(z)$. Substituting (3) and (4) into the "synthesis equation"

$$D(z) = \frac{(1-z^{-1})(1-e^{-T/\tau_s}z^{-1})}{K_s z^{-N-1}(1-z^{-1})(\gamma_3 z^{-1}+\gamma_4)} \cdot \frac{\frac{(z^{-N-1})(\gamma_1 z^{-1}+\gamma_2)}{(1-e^{-T/\tau_s}z^{-1})}}{1 - \frac{(z^{-N-1})(\gamma_1 z^{-1}+\gamma_2)}{(-e^{-T/\lambda}z^{-1})}}$$

which can be simplified to:

$$D(z) = \frac{(\frac{1}{K_s})(\gamma_2 + K_1 z^{-1} - K_2 z^{-2})}{(\gamma_4 + K_3 z^{-1} - K_4 z^{-2} - K_5 z^{-N-1} - K_6 z^{-N-2} - K_7 z^{-N-3})} \quad (5)$$

where:

$K_1 = \gamma_1 - \gamma_2 e^{-T/\tau_s}$

$K_2 = \gamma_1 e^{-T/\tau_s}$

$K_3 = \gamma_3 - \gamma_4 e^{-T/\lambda}$

$K_4 = \gamma_3 e^{-T/\lambda}$

$K_5 = \gamma_2 \gamma_4$

$K_6 = \gamma_3 \gamma_2 + \gamma_4 \gamma_1$

$K_7 = \gamma_3 \gamma_1$

but since $D(z) = \frac{M(z)}{E(z)}$ the above equation can be rewritten as:

$$M(z)(\gamma_4 + K_3 z^{-1} - K_4 z^{-2} - K_5 z^{-N-1} - K_6 z^{-N-2} - K_7 z^{-N-3})$$
$$= E(z)(\frac{1}{K_s})(\gamma_2 + K_1 z^{-1} - K_2 z^{-2})$$

from this the following difference equation is obtained:

$$\gamma_4 m_n + K_3 m_{n-1} - K_4 m_{n-2} - K_5 m_{n-N-1} - K_6 m_{n-N-2} - K_7 m_{n-N-3}$$
$$= (\frac{1}{K_s})(\gamma_2 e_n + K_1 e_{n-1} - K_2 e_{n-2})$$

where:

m_n = controller's output at the present time, n

m_{n-1} = controller's output at one previous time, n-1

m_{n-2} = controller's output at two previous times, n-2

e_n = error at the present time, n

e_{n-1} = error at one previous time, n-1

e_{n-2} = error at two previous times, n-2

Solving for m_n, yields the control algorithm

$$m_n = (\frac{1}{\gamma_4})\left[(\frac{1}{K_s})(\gamma_2 e_{n-1} - K_1 e_{n-1} - K_2 e_{n-2})\right.$$
$$-K_3 m_{n-1} + K_4 m_{n-2} + K_5 m_{n-N-1} + K_6 m_{n-N-2}$$
$$\left. + K_7 m_{n-N-3}\right] \quad (6)$$

SENSITIVITY ANALYSIS

In developing the control algorithm, equation (6) it was necessary to describe the controller system as a first order lag plus dead time. The data for this process model was obtained from a process reaction curve as the one shown in Figure 3. In reality, process reaction curves are seldom, if ever, as smooth as the one presented in the figure. They generally contain superimposed "noise" which can make determination of the process's parameters merely an "educated guess" at the correct values. The analysis presented in this paper investigates how the control of the system is affected when these process parameters are not correctly determined.

The process reaction curve of Figure 3 was generated by simulating on the computer a stirred tank reactor containing no "noise". Therefore, it is assumed that the parameters acquired from this smooth curve represent the most accurate first order approximation of the system possible. It follows then that the system response generated while under control using Dahlin's algorithm (with these actual parameter values) represents the best system control possible from the data available (assuming optimum controller tuning). This "correct" system response is used throughout this chapter as a measure to judge the system responses generated by altering the parameter values of Dahlin's algorithm. The altered parameters simulate incorrect values taken from a process reaction curve with "noise".

In the following paragraphs, the effects of varying the process's parameters, gain, time constant and dead time, used in developing Dahlin's algorithm are discussed in detail. Note that 200 seconds dead time has been added to the actual system (and consequently, also to the dead time for Dahlin's algorithm) in order that this study be relevant to systems which might be controlled by Dahlin's algorithm, i.e., those with a large amount of dead time[3]. Each time a parameter was varied, the control algorithm was optimally tuned, i.e., λ optimally set.

A. Effects of Gain Variations

The results of this sensitivity analysis are shown in Table 1. The integral of the absolute value of the error (IAE) is the criterion used for comparison. The error is defined as the difference between the controller's set point and the controlled variable. From Table 1, it can be seen that the system response deteriorates as the gain deviates from its actual value. Furthermore, the system responses, shown in Figure 4, indicate that the overshoot is the response characteristic most affected by changes in the gain. The overshoot increases as the gain exceeds its actual value and decreases when it is less.

B. Effects of Time Constant and Dead Time Variations

Reproduced in Figures 5 and 6 are the system response curves generated by varying the time constant and dead time, respectively, from their actual values. These figures indicate that the system's response deteriorates whenever the dead time or time constant is reduced below their actual values. These findings are substantiated by the IAE values corresponding to these curves as shown in Table 1.

This type of system behavior can be explained by recalling that Dahlin's controller output is a function of the current and past errors and controller's past outputs. As the design system parameters (time constant and dead time) are reduced, they change Dahlin's algorithm to a function which will give optimum control to a "faster" system than the real system being controlled. This results in an algorithm controlling a system which cannot, due to its inherent slower response, keep up with the designed algorithm: thus, as the process parameters for Dahlin's algorithm are reduced the system response to a step change in set point deteriorates. To summarize, designing Dahlin's controller with dead time and time constants smaller than the actual values results in a controller too "fast" for the controlled system.

It is interesting to note that as either the dead time or time constant is increased above its actual value, the system response first improves and then deteriorates, according to the IAE values of Table 1. This type of system behavior indicates that the parameters at +10% (for the time constant and dead time only), when used in conjunction with the first order approximation of the system being controlled, represent the best possible configuration of parameters to minimize the IAE. There are two possible explanations for these results. The first explanation is due to the fact that Dahlin's algorithm is "slowed down" by increases in either the process dead time or time constant used in the development of the algorithm. This allows the "faster" control loop to respond more in line with the algorithm's design. Further increases in the dead time or time constant results in improving the control loop's ability to "follow" the algorithm, however, the overall effect is a deterioration in the control of the process.

The second possible explanation for the control improvement is that these new parameters used in development of the control algorithm could be a better first order approximation to the controlled system than the one previously used.

While varying the dead time of the system for the sensitivity analysis presented in this chapter, it was found that an instability exists with Dahlin's algorithm. A reference to this possible instability has already been mentioned in the literature[4]. It was found that the only difference between a stable run and an unstable one was that K_1 of equation (3) was positive for unstable runs and negative for stable ones. The only parameter difference between a stable run and an unstable run was the dead time used in Dahlin's algorithm.

Through further analysis, it was determined that the algorithm approaches instability as K_1 (a negative value) approaches zero and is always unstable whenever $K_1 \geq 0.0$. Until further research can be performed in this area, it is suggested that whenever Dahlin's algorithm is used with a process, the system parameters and the sampling time be specified in such a manner that K_1 is always negative. A way to make sure that this condition, K_1 negative, is satisfied is to make the process dead time, θ_s, and the one specified for the closed loop response, θ_L, an integer number of sampling times, T.

C. Comparison of Results

By comparing Figures 4, 5 and 6, and by analyzing the figures in Table 1, it is apparent that of the three parameters (gain, time constant and dead time), the gain variations affected the system response the least while the dead time variations altered the responses the most. Table 1 further suggests that the integral of the absolute value of the error might be further reduced by using the actual gain along with time constant and dead time values from the 10% levels. It is felt, however, that such a reduction would not affect appreciably the settling time of the process.

To investigate the effect on the system of varying all three parameters simultaneously, two runs were made by varying all of the parameters to the $\pm 20\%$ levels. The results of these runs are shown in Figure 7 and at the bottom of Table 1. They indicate that it is by far better to overestimate the process parameters for Dahlin's algorithm than it is to underestimate them. In fact, the IAE figures indicate that the system response deteriorates by 42.14% when the parameters are decreased to the -20% level and deteriorates by only 5.2% when the parameters are increased by 20% above their actual values. It can therefore be concluded that, when taking parameters from a process reaction curve with "noise" for Dahlin's algorithm, it would be better to take the values from the high side of the ranges obtained.

The discussions and conclusions presented in this paper are limited to the system under study. They were presented to yield insight into the use of Dahlin's algorithm. Any application of the conclusions from this paper to systems other than the one presented in Appendix A should therefore be limited to non-specific trends keeping in mind their origin and thus their limitations.

NOMENCLATURE

$e(t)$ = controller's error = set point = measured variable, psig

e_n = present error for discrete controller

Kc = controller's gain, psig/psig

$m(t)$ = continuous controller's output, psig

m_n = present output from discrete controller

m_{n-1} = previous output from discrete controller

m_o = controller's bias value, psig

T = sampling time, sec

τ_I = reset time, min/repeat

τ_D = rate time, minutes

τ_s = actual process time constant from process reaction curve, sec

θ_s = actual process dead time from process reaction curve, sec

K_s = actual process gain from process reaction curve psi/psi

REFERENCES

1. Dahlin, E. B., "Design and Tuning Digital Controllers, Part I," <u>Instrument and Control Systems</u>, Vol. 41, No. 6, 1968.

2. Smith, C. L., "Digital Computer Process Control", Intex Educational Publishers, Scranton, Pa. 1972

3. Condon, B. T., Smith, C. A., "A Comparison of Controller Algorithms As Applied to a Chemical Reactor", Instrument Society of America Meeting, Anaheim, California, 1977

4. Higham, J. D., "Single Term Control of First and Second Order Processes With Dead Time", <u>Control</u>, February 1968.

APPENDIX A: REACTOR MODELING, SPECIFICATIONS AND SIMULATION

This appendix presents the description of the stirred tank chemical reactor used in this work. Specifically, the modeling, specification and computer simulation of the reactor are presented. The response of this reactor is quite nonlinear arising mainly from the Arrhenius term in the kinetic expression.

The reaction being controlled is an irreversible exothermic reaction of the type:

$$A \xrightarrow{k} B$$

with first oreer kinetics taking place. The reaction rate is given by:

$$r_A = k_0 \exp[E/R(T + 460.)]C_A \quad (A.1)$$

where:

r_A = reaction rate, moles of formed ft^3-hr

k_0 = frequency factor, 7.08×10^{10} hr^{-1}

E = activation energy, 30000 Btu/mole

R = gas constant, 1.987 Btu/mole-^oK

T = temperature, ^oF

C_A = concentration of A in the reactor, moles/ft^3

The mass balance on the reactor contents, assuming uniform mixing and constant density, is:

$$\rho F_0 - \rho F = \rho A \frac{dh}{dt} \quad (A.2)$$

where:

F_0 = inlet flow, 2000 ft^3/hr

F = outlet flow, ft^3/hr

A = reactor's inside cross sectional area, 5.0 ft^2

h = level in reactor, ft.

ρ = liquid density, assumed contstant, 50 lb_m/ft^3

The mass balance on component A is:

$$F_0 C_{A0} - F C_A - Ah\, r_A = A\frac{d(hC_A)}{dt} \quad (A.3)$$

where:

C_{A0} = inlet concentration of A, 0.4 moles/ft^3

Since the reactor is assumed to be well mixed, the reactor's content concentration equals the outlet concentration.

The heat balance on the reactor's contents is:

$$F_0 \rho C_p (T_0 - T_R) - \Delta H_r Ah r_A - h_i A_i (T - T_m)$$
$$- F\rho C_p (T - T_R) = \rho A C_p \frac{d[h(T-T_R)]}{dt} \quad (A.4)$$

where:

C_p = liquid heat capacity, assumed constant, 0.75 Btu/lb_m °F

T_0 = reactant's inlet temperature, 250 °F

T_R = reference temperature, 32.0 °F

ΔH_r = heat of reaction, -1500 Btu/mole

h_i = inside film heat transfer coefficient, assumed constant, 160 Btu/hr-ft^2-°F

A_i = inside heat transfer area, assumed constant, 250 ft^2

T_m = metal wall temperature, °F

T = reactor's content temperature, °F

The heat balance on the cooling water jacket is:

$$F_J \rho_J C_{pJ}(T_{J_0} - T_J) + h_0 A_0 (T_m - T_J)$$
$$= V_J \rho_J C_{pJ} \frac{d(T_J)}{dT} \quad (A.6)$$

where:

F_J = volumetric flow rate of water through the cooling jacket, ft^3/hr

ρ_J = water density, assumed constant, 62.4 lb_m-°F

C_{pJ} = heat capacity of water, 1.0 Btu/lb_m-°F

T_{J_0} = inlet water temperature, 70 °F

V_J = jacket volume, 15.0 ft^3

The flow rate out of the reactor tank is calculated according to the following equation for flow through a valve:

$$F = 8.02\, C_v\, v_p \sqrt{\frac{\Delta p}{G}} \quad (A.7)$$

where:

C_v = valve coefficient, 239.4

v_p = valve position

Δp = pressure drop over valve, psig assuming that the reactor is open to the atmosphere and that the valve discharges to the atmosphere, i.e.:

$$\Delta p = \frac{\rho g h}{g_c} = \rho h/144 \text{ psia}$$

G = specific gravity of reactor's contents, 0.8

The cooling water flow rate (F_J) is calculated using the same equation:

$$F_J = 8.02\, C_{vJ}\, v_{pJ} \sqrt{\frac{\Delta p_J}{G_J}} \quad (A.8)$$

where:

C_{vJ} = water jacket valve coefficient, 34.8

v_{pJ} = valve position of water jacket valve

Δp_J = pressure drop across the water jacket valve, 30 psi

G_J = water specific gravity, 1.0

In addition to the above equations that fully describe the reactor, some other descriptive equations are necessary to complete the control loop.

In order to measure and transmit the outlet reactor's concentration to the concentration controller, a concentration transmitter is used. This transmitter is calibrated for a range of 0 - 1.5 moles/ft^3. Since this is a pneumatic transmitter, its output range is 3 - 15 psig. This calibration allows us to obtain the transmitter gain, K_{CT}:

$$K_{CT} = \frac{(15-3)\text{ psig}}{(1.5-0)\text{ moles/}ft^3} = 8.0 \text{ psig/moles/}ft^3$$

The descriptive equation for this transmitter is:

$$C_{AT} = 3.0 + K_{CT}(C_A) \quad (A.9)$$

where:

C_{AT} = concentration of component A as transmitted by the transmitter, psig

The level transmitter used to measure the liquid in the reactor has a calibration of 0 - 20 ft for an output of 3 - 15 psig. Therefore, this transmitter gain, K_{LT}, is:

$$K_{LT} = \frac{(15-3) \text{ psig}}{(20-0) \text{ ft}} = 0.6 \text{ psig/ft}$$

The descriptive equation is:

$$h_T = 3.0 + K_{LT}(h)$$

where:

h_T = transmitted level, psig

The cooling water valve actuator can be described as:

$$F_J vp = K_v(m_c) - 0.25 \quad (A.11)$$

where:

$F_J vp$ = outlet from actuator to the valve, valve position

m_c = input to actuator from concentration controller, psig

K_v = actuator's gain, valve position/psig

The liquid level valve actuator is also described as follows:

$$F_L vp = K_v(m_h) - 0.25 \quad (A.12)$$

where:

$F_L vp$ = output from the actuator to the valve, valve position

m_h = input to actuator from level controller, psig.

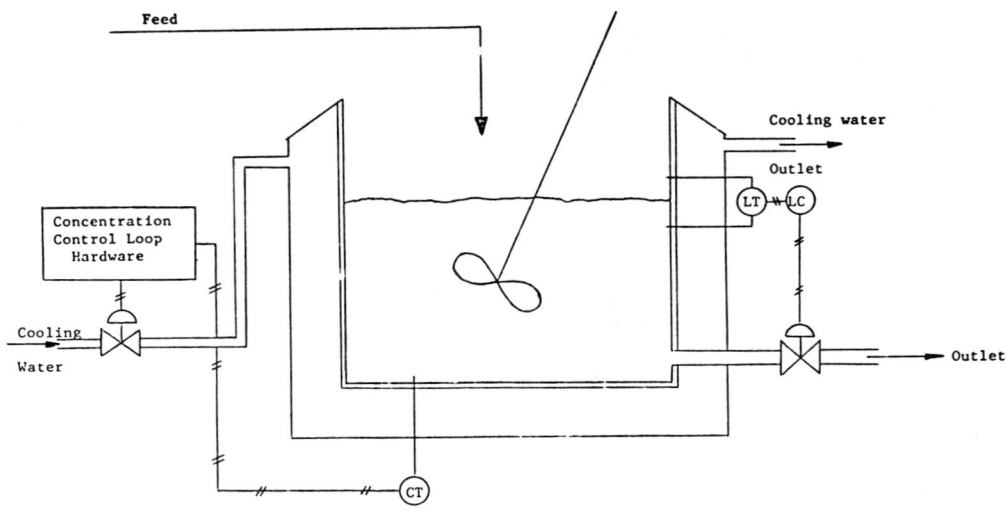

Figure 1   Stirred tank reactor with control loops.

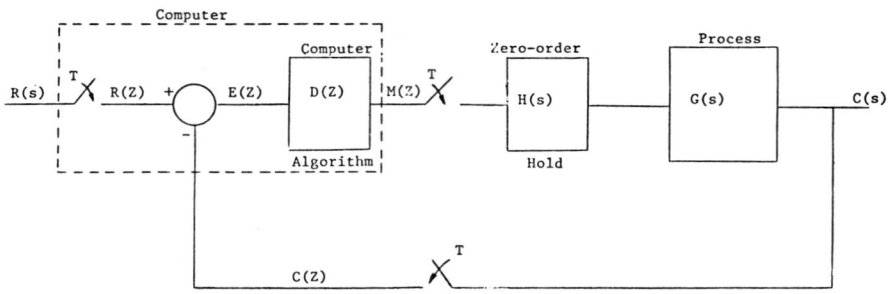

Figure 2   Reduced Block Diagram of the Direct Digital Control System.

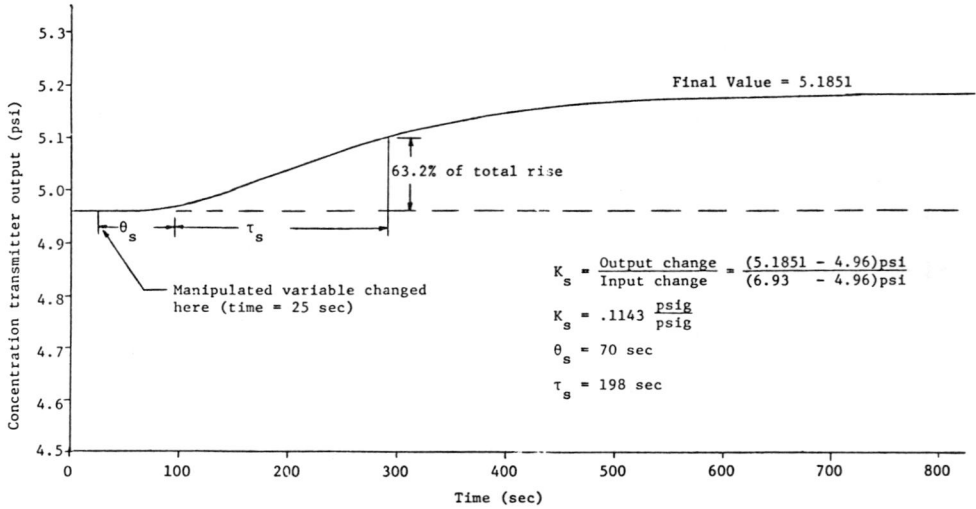

Figure 3   Process reaction curve for reactor (valve actuator output changed from 4.96 psia to 6.93 psia)

% Deviation from Actual Value	-20%	-10%	0.0%	+10%	+20%
Error Criteria	IAE	IAE	IAE	IAE	IAE
Gain	75.73	73.586	73.143	73.29	74.22
Time Constant	80.095	76.563	73.143	71.645	72.138
Dead Time	88.996	80.820	73.143	70.066	70.777
All Parameters	104.14		73.143		76.942

Table 1   SENSITIVITY ANALYSIS RESULTS MATRIX.

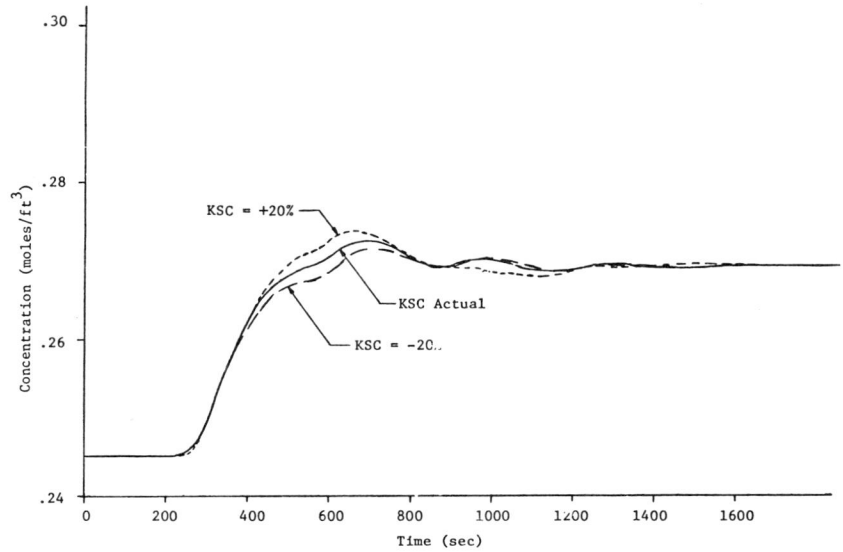

Figure 4   Concentration Response as Dahlin's Gain Parameter Changes.

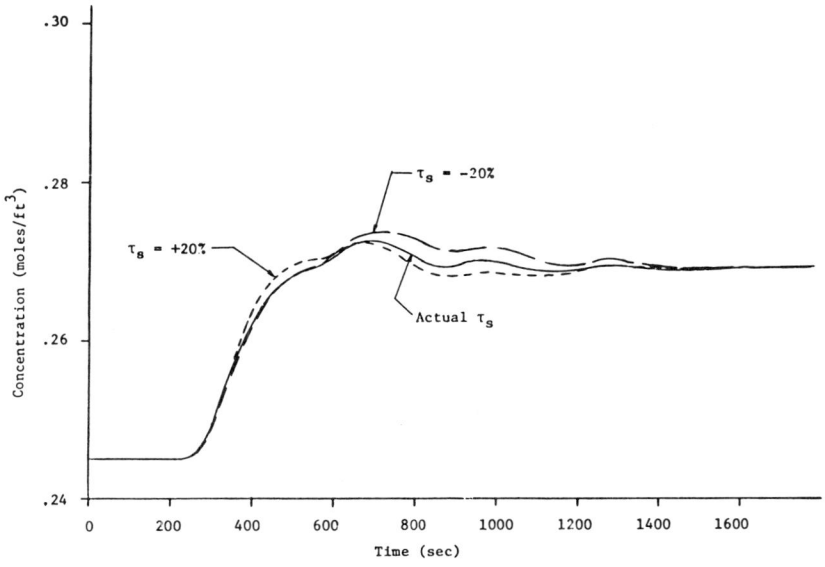

Figure 5   Effect on Control Loop Response by Varying the Time Constant From Its Actual Value.

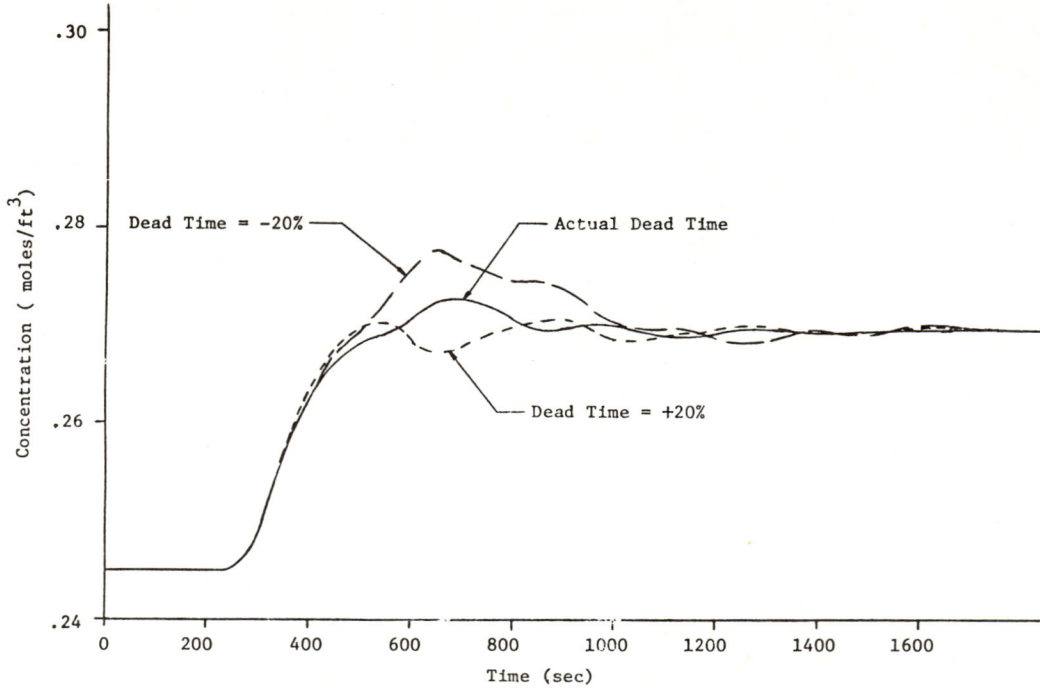

Figure 6   Effect on Control Loop Response by Varying the Dead Time From Its Actual Value.

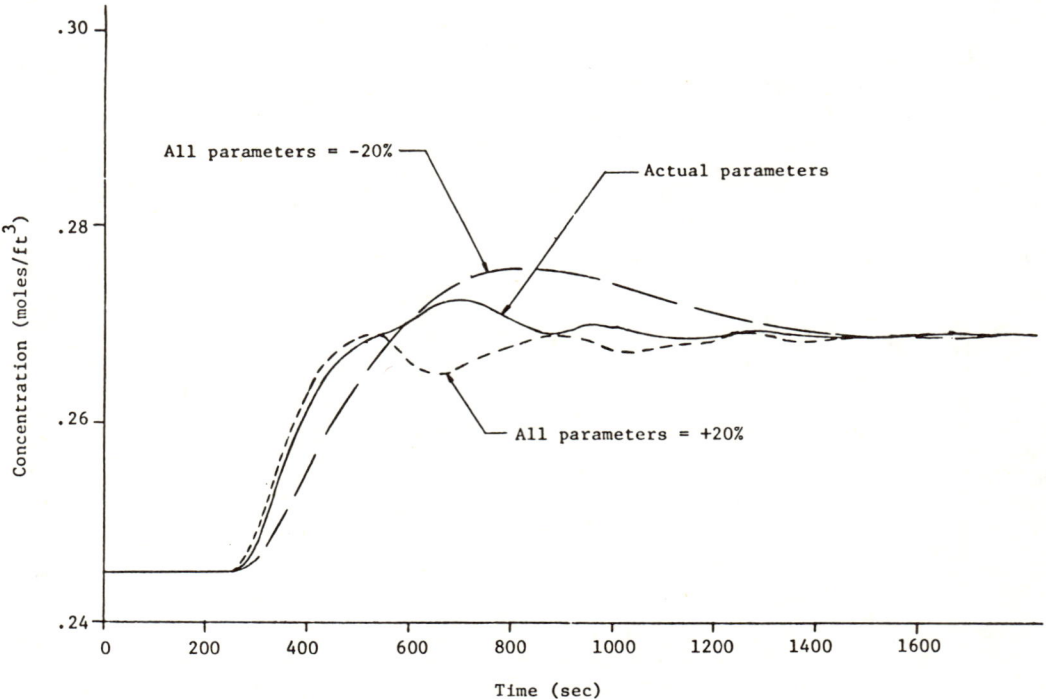

Figure 7   Effect on Control Loop Response by Varying All Three Parameters from Their Actual Values.

©ISA, 1977
ISBN 87664-363-2

COMPARISON OF TUNING METHODS FOR TEMPERATURE
CONTROL OF A CHEMICAL REACTOR

Jacob Martin, Jr., A. B. Corripio and C. L. Smith
Department of Chemical Engineering
Louisiana State University
Baton Rouge, Louisiana 70803

ABSTRACT

This paper presents the results of applying various controller tuning methods to the feedback temperature control of a simulated chemical reactor. Specifically the minimum error-integral correlations presented by Lopez and by Rovira are compared with the Ziegler-Nichols method and the loop-compensation method previously presented by these authors. Based on speed of response and control stability the best all-around techniques are shown to be Rovira's set-point tuning and the loop-compensation method. The closed-loop responses of the reactor temperature to changes in set-point and load are presented.

INTRODUCTION

The modern control engineer finds that his task of designing feedback control systems involves the specification of the variables to be measured, the streams to be manipulated, the sensors and transmitters to do the measuring, the control valve, complete with actuator and positioner if so required, and the feedback controller. A key step in the complete operation is the tuning of the controller parameters to the dynamic characteristics of the process. The success or failure of the design process may depend on it.

The process control literature of the last four decades contains a number of tuning methods and correlation formulas for feedback controllers. Of these, we have chosen to consider in this paper the minimum error integral correlations of Rovira [1] for set-point inputs and of Lopez [2] for disturbance inputs, the pioneer method of Ziegler and Nichols [3] and the loop-compensation method proposed by Martin et al [4]. These methods are chosen because they are intended to apply in general to the most common processes. Other methods have been presented in the literature that apply to more specific processes such as the control of pH, centrifugal compressors, liquid level, etc. The methods studied here would in no way compete with the more specific methods and correlations for these process loops.

TEMPERATURE CONTROL OF A CHEMICAL REACTOR

In order to study the performance of the controller to a more realistic plant than a first or second-order differential equation, the temperature control of a continuous stirred tank chemical reactor was chosen. The reactor, which is sketched in Figure 1, was simulated on an analog computer with enough detail as to include its most important non-linear dynamic characteristics. A detailed description of the mathematical model has been given in a previous publication [5].

The temperature in the reactor is controlled at the desired set-point by manipulation of the rate of cooling water to the jacket. Step changes in temperature set-point and load are applied to observe the response of the controller tuned by the methods under consideration. The load input is the rate of reactant feed.

The nonlinear behavior of the loop is illustrated in Figure 2 by a steady-state plot of temperature - the control variable - versus cooling water rate - the manipulated variable. The slope of this line is proportional to the process gain and is shown to increase as the cooling water rate decreases.

MODELS FOR CONTROLLER TUNING

All of the tuning methods considered here make use of correlations which are based on the parameters of either a first-order lag plus dead-time (transportation lag or time delay) or a second-order lag plus dead-time. These parameters are most commonly obtained from a process reaction curve of the process. Such a curve is the open-loop response of the controlled variable to a step-change in manipulated variable. For the reactor, a first-order plus dead-time model was obtained by a method proposed by Miller [6] and the parameters obtained are as follows:

$$\tau \frac{dT}{dt} + T(t) = KW_c(t-t_o) \qquad (1)$$

K = -0.0333 °F/(lbs/min)
τ = 13.95 min
t_o = 2.5 min

The parameters of the second-order lag plus dead-time were determined by Sten's method [7], and are as follows:

$$\frac{d^2T}{dt^2} + b\frac{dT}{dt} + cT(t) = cKW_c(t-t_o) \qquad (2)$$

$K = -0.0333 \, °F/(lbs/min)$
$b = 0.34 \, min^{-1}$
$c = 0.024 \, min^{-2}$
$t_o = 0.8 \, min$

These parameters correspond to an overdamped system with time constants of 10.1 and 4.1 min.

The process reaction curve was obtained by a step reduction in cooling water rate of 120 lbs/min. Although the parameters vary with the magnitude and direction of the step, we felt that it would be unrealistic in an industrial situation to perform multiple tests to obtain an average. A single process reaction curve is difficult enough to get.

TUNING METHODS

Rovira [1] and Lopez [2] based their tuning correlations on the first-order plus dead-time model of equation 1. They used parameter search techniques to determine the controller parameters that result in a minimum of the following error functions:

$$IAE = \int_0^\infty |e| \, dt$$

$$ITAE = \int_0^\infty |e| t \, dt$$

where $|e|$ is the absolute value of the error or instantaneous difference between the controlled variable and the set-point. Lopez also considered the integral of the squared error (ISE) but the resulting tuning parameters gave highly oscillatory responses.

While Lopez minimized the error integrals for step changes in disturbance - assuming the process dynamics to disturbance were identical to the dynamics to the controller output signal -, Rovira considered step changes in set-point, a more demanding case. This resulted in more conservative tuning parameters.

Although Ziegler and Nichols [3] did not base their tuning correlations on the parameters of equation 1, the graphically defined parameters from the process reaction curve can be converted to those of the first-order plus dead-time model. They obtained their correlations by empirical methods and based them on a quarter-decay ratio. Unfortunately, the quarter-decay ratio is a more oscillatory response than is usually acceptable in an industrial environment.

These authors' loop-compensation approach [4] consists of looking at the controller as a dynamic compensator for the other major component of the loop: the process. In his approach the integral time is considered a zero of the controller transfer function that is used to compensate for the longest time constant of the process or dominant pole. The derivative time is considered as a second zero that compensates for the second longest process time constant. With the two time parameters thus determined to insure fast loop response, the gain can then be adjusted to meet any specified response criteria. Gain correlations were obtained for 5% overshoot on the response to a step change in set-point.

Martin's approach requires the first-order plus dead-time model parameters to tune a proportional-integral (PI) controller, and the second-order plus dead time parameters to tune a proportional-integral-derivative controller (PID).

Tuning the Temperature Controller

Using the parameters obtained from the process reaction curve for the reactor, the PI controller parameters computed from the various correlations are shown in Table I. Note that the parameters from the Rovira correlation are essentially the same - within control parameter accuracy - as those from the loop-compensation approach for 5% overshoot, while the Lopez and Ziegler-Nichols correlations result in tighter control parameters. The equation for the PI controller is given by:

$$m = m_o + K_c \left(e + \frac{1}{T_i} \int e \, dt\right) \qquad (3)$$

where $e = $ (set-point) $- T$

The parameters for the PID controller are given in Table II. Note that the integral and derivative times for the loop-compensation method are of the order of magnitude of the process time constants. Again the Lopez and Ziegler-Nichols correlations result in higher gains and shorter integral time. The parameters of Table II are for the following PID controller equation:

$$m = m_o + K_c \left(e + \frac{1}{T_i} \int e \, dt + T_d \frac{de}{dt}\right) \qquad (4)$$

COMPARATIVE SET-POINT RESPONSES

The responses of the PI controller to a 4 °F rise in set-point are given in Figure 3. The figure illustrates the closeness of the response between the Rovira and loop-compensation techniques. Lopez' tuning is not shown because it is not intended for set-point changes. The quarter-decay Ziegler-Nichols response is obviously too oscillatory. Note that the loop-compensation response exceeds the 5% overshoot for which it was designed. This is because of the nonlinear nature of the reactor which, at this higher temperature, exhibits a higher gain than measured by the process reaction curve. The nonlinear effect is evident in Figure 4, which shows the response of the PI controller to a 4 °F drop in set-point. At the lower temperature the reactor gain is lower.

The corresponding responses of the PID controller tuned by the various correlations are shown in Figures 5 and 6 for a 4 °F rise and drop in set-point, respectively. The Rovira IAE tuning has a faster rise-time than loop-compensation tuning. It must be kept in mind, however, that the gain of the controller can be field adjusted under the loop-compensation concept, since it is determined independently of the integral and derivative times.

RESPONSES TO LOAD INPUT

The PI controller responses to a 20% decrease in reactant feed rate are shown in Figure 7. In this case the Lopez tuning is included since it is intended for load changes. Rovira's tuning is essentially the same as loop-compensation tuning for the PI controller as mentioned earlier. Note the oscillatory behavior of the Ziegler-Nichols and Lopez responses. Again it appears that the loop-compensation gain could be adjusted upwards to speed up its response. The corresponding PID controller responses are shown in Figure 8. The Lopez and Ziegler-Nichols responses are less oscillatory in this case. Again, loop-compensation is rather conservative, and seems that it could stand a higher gain.

SUMMARY AND CONCLUSIONS

Tuning techniques based on simple models of the process have been shown to result in very satisfactory responses when applied to the temperature control of a nonlinear chemical reactor. The tuning methods of Rovira and loop-compensation produced the best responses to set-point changes in terms of stability and speed of response. In general, set-point changes are more demanding in controller performance than load changes, and are equivalent to load changes when the process dynamics to disturbance inputs are much faster than its dynamics to the manipulated variable.

Although conservative in terms of responses to load changes, the fact that the loop-compensation concept decouples the adjustment of the gain from that of the time parameters allows the operator to obtain any desired response criteria by adjustment of only one knob. This is not the case for the other methods for which all of the controller parameters are interrelated.

ACKNOWLEDGEMENT

This work was sponsored by a grant from the Air Force Office of Scientific Research, Air Force Systems Command, USAF, Contract No. 74-2580.

NOTATION

b	Second-order model damping parameter
c	Second-order model frequency parameter
e	Controller error
IAE	Integral of the absolute value of the error
ITAE	Integral of time averaged absolute value of the error
K	Model gain
K_c	Controller gain
m	Controller output signal
m_o	Initial controller output signal
t	Time
t_o	Model dead-time
T	Temperature of the reacting fluid
T_d	Controller derivative time
T_i	Controller integral time
W	Mass rate of feed (load input)
W_c	Mass rate of cooling water (manipulated variable)
τ	Model time constant

REFERENCES

[1] Rovira, A. A., P. W. Murrill, and C. L. Smith, "Tuning Controllers for Setpoint Changes", *Instruments and Control Systems*, Vol. 42, No. 12, December 1969, p. 67.

[2] Lopez, A. M. et al., "Controller Tuning Relationships Based on Integral Performance Criteria", *Instrumentation Technology*, Vol. 14, No. 11, November 1967, p. 57.

[3] Ziegler, J. G. and N. B. Nichols, "Optimum Settings for Automatic Controllers", *Trans. ASME*, Vol. 64, No. 11, November 1942, p. 759.

[4] Martin, Jacob, Jr., Corripio, A. B. and Smith, C. L., "Controller Tuning from Simple Process Models", *Inst. Tech.*, Vol. 22, No. 12, Dec. 1975, pp. 39-44.

[5] Corripio, A. B. and C. L. Smith, "Computer Simulation to Evaluate Control Strategies", *Instruments and Control Systems*, Vol. 44, No. 1, January, 1971, p. 87.

[6] Miller, J. A., et al., "A Comparison of Controller Tuning Techniques", *Control Engineering*, Vol. 14, No. 12, December 1967, p. 72.

[7] Sten, J. W., "Evaluating Second-Order Parameters", *Instrumentation Technology*, Vol. 17, No. 9, September 1970, pp. 39-41.

TABLE I
PI CONTROLLER TUNING PARAMETERS - BACKMIX REACTOR

Techniques	K_c (lbs/min/°F)	T_i min
Loop-Compensation (5% overshoot)	-0.584	13.95
Rovira (Set-Point) (IAE)	-0.666	14.50
Lopez (Load) (IAE)	-1.072	6.805
Ziegler-Nichols (quarter decay)	-1.003	8.32

TABLE II
PID CONTROL TUNING PARAMETERS - BACKMIX REACTOR

Techniques	K_c lbs/min/°F	T_i min.	T_d min
Loop-Compensation (5% overshoot)	-0.404	14.18	2.94
Rovira (Set-Point) (IAE)	-0.968	19.46	1.01
Lopez (Load) (IAE)	-1.398	4.383	0.952
Ziegler-Nichols (quarter-decay)	-1.116	5.00	1.25

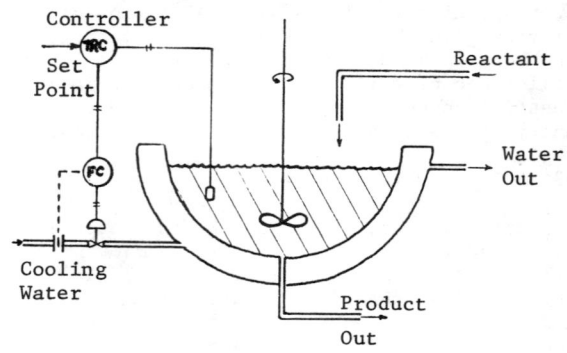

Figure 1. Reactor temperature control scheme

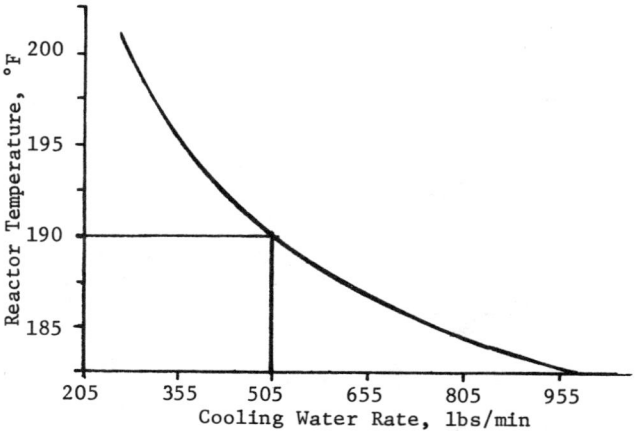

Figure 2. Reactor temperature versus cooling water rate at steady-state

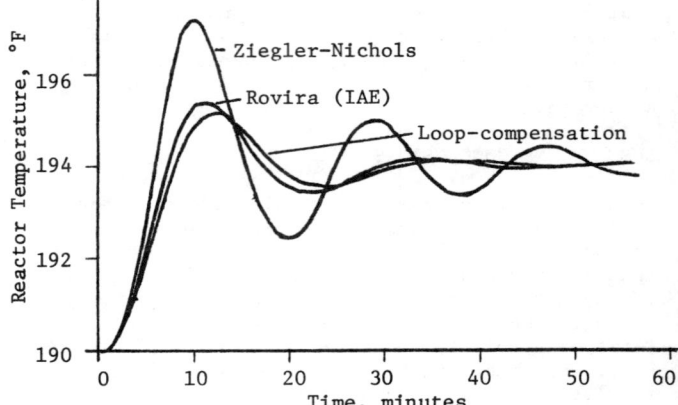

Figure 3. Closed-loop response of PI controller for a 4°F rise in set-point

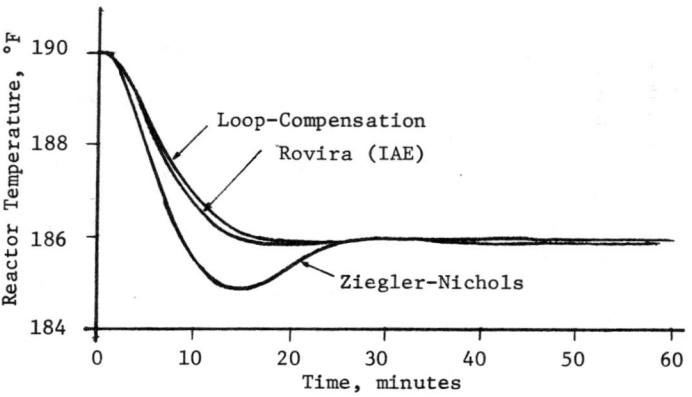

Figure 4. Closed-loop response of PI controller for a 4°F drop in set-point

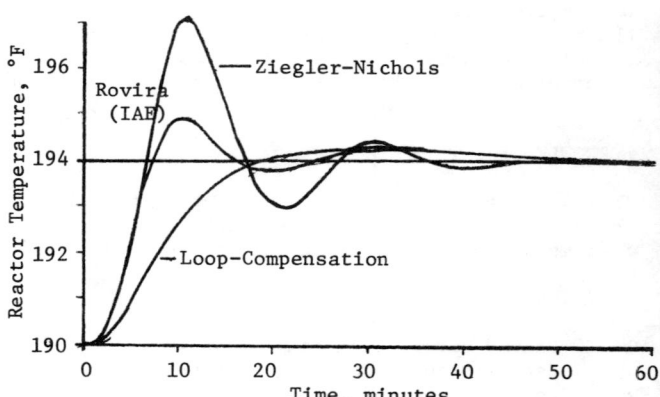

Figure 5. Closed-loop response of PID controller for a 4°F rise in set-point

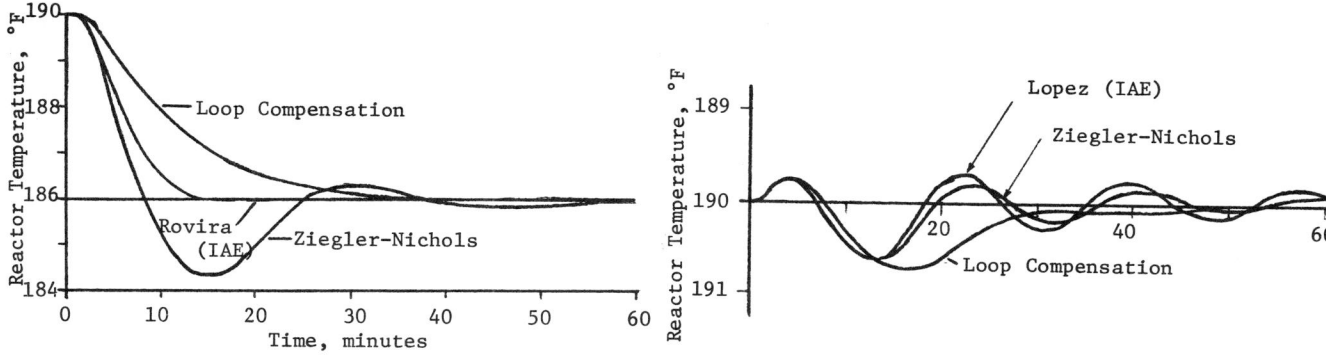

Figure 6. Closed-loop response of PID controller for a 4°F drop in set-point

Figure 7. Closed-loop response of PI controller for a 20% decrease in reactant feed rate

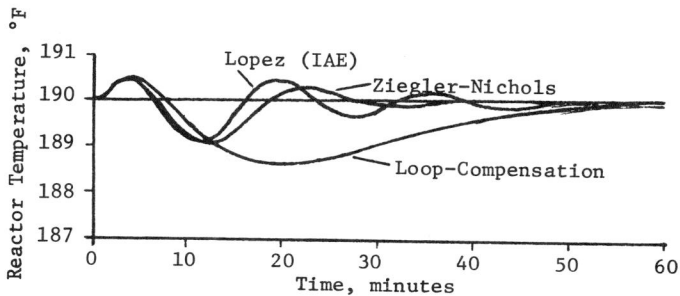

Figure 8. Closed-loop response of PID controller for a 20% decrease in reactant feed rate

©ISA, 1977
ISBN 87664-363-2

EMPIRICAL SECOND ORDER NONLINEAR PROCESS MODEL

DEVELOPMENT AND APPLICATION

Carlos A. Smith
University of South Florida
Tampa, Florida 33620

Frank R. Groves, Jr.
Louisiana State University
Baton Rouge, Louisiana 70803

INTRODUCTION

In the recent literature, a paper [1] has described the development and application of two empirical first order nonlinear process models. These models were shown to describe processes more accurately over a wider range of the state variables than the linear models more often used. The application of the nonlinear models to implement an optimal control scheme was also shown. Another paper [2] has presented the application of the first order nonlinear models to a dead time compensation scheme for an oil heater.

The present paper describes the development of an empirical second order nonlinear process model. As with the first order models, this model is mathematically simple. Its parameters are readily determined from several step function response curves from the process. The model was tested by application to a simulated stirred tank chemical reactor. It was used successfully to implement a time optimal control scheme on the simulated reactor. Product concentration from the reactor was controlled by manipulating water flow rate to the reactor cooling coil.

Time optimal control consists of developing a control law to drive the reactor's outlet concentration from its nominal steady state to a new level in minimum time. Two control laws were derived, one using the nonlinear model and another using a linear model. A comparison of the two models, when applied to the reactor showed the superiority of the nonlinear model.

DEVELOPMENT OF THE MODEL

Mathematically, a second order linear model is of the following form:

$$\tau_1 \tau_2 \frac{d^2 C(t)}{dt^2} + (\tau_1 + \tau_2) \frac{dC(t)}{dt} + C(t) = KM(t) \quad (1)$$

The variable $C(t)$ is the state variable in perturbed form. The variable $M(t)$ is the disturbance in perturbed form. K, τ_1 and τ_2 are the model's parameters. The development of an empirical nonlinear second order model is similar to the development of the empirical first order nonlinear model. From this development it was found that, for the cases studied, the parameter K, gain, is almost linearly related to the final value of the state variable, C_{ss}, obtained by applying a step change disturbance to the process. To review, let us go over how this was done. The idea is to disturb the process in a step manner, that is, by changing the value of $M(t)$ from its nominal value to a new level, $M1$. From the response of the process we can obtain a process reaction curve. Figure 1 shows a typical process reaction curve. The ordinate is the value of the actual state variable, $c(t)$, not the perturbed variable, $C(t)$. From this curve we can obtain a value of K by dividing the change in the value of the state variable, Δc_1, by the magnitude of the disturbance, Δm_1:

$$K = \frac{\Delta c_1}{\Delta m_1}$$

Remembering that a perturbed variable equals the actual variable minus its nominal steady state, we can express the above as:

$$C_1 = \Delta c_1$$
$$M_1 = \Delta m_1$$

therefore

$$K = \frac{C_1}{M_1}$$

This gives us a point on the graph of K versus C_{ss}. The value of C_{ss} is the new steady state, C_1, of the process. We continue doing this for three or more values of the disturbance. At this point we have enough data points to obtain a relationship. As mentioned above, for the cases studied, a linear relationship provides a satisfactory fit, that is:

$$K = a3 + b3\, C(t) \quad (2)$$

Murrill [3] explains how to obtain the other two parameters, τ_1 and τ_2, for the linear model. In developing the nonlinear model it was noticed that the terms $\tau_1 \tau_2$ and $(\tau_1 + \tau_2)$ of the linear model behave in a similar manner to the parameter K. That is, a relation between $\tau_1 \tau_2$ and $C(t)$ and between

$\tau_1 + \tau_2$ and $C(t)$ can also be obtained. The way to do this is to evaluate, as is done for the linear model, for each disturbance the parameters τ_1 and τ_2. From these values we can obtain the terms $\tau_1 \tau_2$ and $\tau_1 + \tau_2$. Each of these terms are then plotted versus C_{ss} as was done with the parameter K. From each plot we can then obtain a linear relationship, that is,

$$\tau_1 \tau_2 = a1 + b1\ C(t) \quad (3)$$

$$\tau_1 + \tau_2 = a2 + b2\ C(t) \quad (4)$$

Substituting equations (2), (3) and (4) into equation (1) the model becomes:

$$(a1+b1\ C(t)) \frac{d^2C(t)}{dt^2} + (a2+b2\ C(t)) \frac{dC(t)}{dt} + C(t)$$

$$= (a3+b3\ C(t))\ M(t) \quad (5)$$

When the terms K, $\tau_1 \tau_2$ and $\tau_1 + \tau_2$ are plotted versus C_{ss} the data will not generally fall exactly on a straight line. However, the straight line gives a representation of the trend of the data. Determination of a3 and b3 in this way tends to make the model represent the steady state response of the system over the range of $C(t)$ covered by the test. The parameters a1, b1, a2, b2 cause the terms $\tau_1 \tau_2$ and $\tau_1 + \tau_2$ to be adjusted automatically to a value appropriate to the range of $C(t)$ in a particular test.

EVALUATION OF THE NONLINEAR MODEL FOR A CHEMICAL REACTOR

The nonlinear model, equation (5), was implemented for a system consisting of a simulated stirred tank chemical reactor. The reactor is the same as that described by Koppel [4] and was chosen because it is a highly nonlinear system. The reactor is an exothermic first order irreversible reaction of the type A → B. Heat is removed from the stirred tank by means of cooling coil. The state variable $c(t)$ is the concentration of reactant A in the product stream from the tank. The manipulated variable is the cooling water flow rate, F_c. The simulation of the reactor consists of three ordinary differential equations - a material balance on the stirred tank, and an energy balance on the cooling coil. The equations and values of the parameters for the simulation are given in Appendix I. The equations were solved on an XDS Sigma 5 digital computer using the Runge-Kutta method.

For the purpose of model evaluation the reactor was subjected to step disturbance starting at a steady state with $c_{ss} = 0.162 \times 10^{-3}$ g-mole/cc and $m_{ss} = 5.13$ cc/sec. Disturbances in cooling water flow rate of $M_1 = -2.0$ cc/sec, $M_2 = 1.0$ cc/sec and $M_3 = 3.0$ cc/sec were imposed. Figure 2 shows the reactor's response. These responses are shown in actual state variables $c(t)$. Table I shows values of K, τ_1, τ_2 derived from these responses. The values of K, τ_1 and τ_2 were obtained following Murrill [3]. Figures 3, 4, and 5 show the relationships K vs. C_{ss}, $\tau_1 \tau_2$ vs. C_{ss} and $(\tau_1 + \tau_2)$ vs. C_{ss} respectively. From these graphs we obtain:

$$K = 0.613 \times 10^{-4} + 0.0891\ C(t) \quad (6)$$

$$\tau_1 \tau_2 = 870.42 + 5383.5 \times 10^3\ C(t) \quad (7)$$

$$\tau_1 + \tau_2 = 66.35 + 172.9 \times 10^3\ C(t) \quad (8)$$

Therefore, the model is:

$$(870.42 + 5383.5 \times 10^3\ C(t)) \frac{d^2C(t)}{dt^2}$$

$$+ (66.35 + 172.9 \times 10^3\ C(t)) \frac{dC(t)}{dt} + C(t)$$

$$= (0.613 \times 10^{-4} + 0.0891\ C(t))\ M(t) \quad (9)$$

Let us demonstrate the advantage of the nonlinear model over the linear model. This is done by comparing the responses of the linear and nonlinear models with that of the simulated reactor. The linear model of the reactor is:

$$1143.45 \frac{d^2C(t)}{dt^2} + 91.85 \frac{dC(t)}{dt} + C(t)$$

$$= 0.672 \times 10^{-4}\ M(t) \quad (10)$$

This model is obtained from the response to a 1 cc/sec disturbance. Figures 6 and 7 show the responses of the models and of the simulated reactor when step disturbances of 2.17 cc/sec and 3.0 cc/sec are applied. The nonlinear model describes the response of the reactor much better than the linear model. These two figures demonstrate that the nonlinear model describes the process more accurately over a wider range of the state variable than the linear model. For the disturbance of 1.0 cc/sec, the linear model describes the process as well as the nonlinear model. This is so because the linear model was evaluated based on this disturbance. Before continuing let us note one important point. The disturbances of 2.17 cc/sec and 3.0 cc/sec represent changes of 38.9% and 58.4% respectively in cooling water flow, the forcing function.

Next, let us compare the models when another kind of disturbance is applied. This new disturbance is ramp, that is, the forcing function is increased at a constant rate:

$M(t) = 0.0125 *$ time cc/sec

This continues up to $M(t) = 5$ cc/sec. Then this value is kept constant. Figure 8 shows the three responses. Again, the nonlinear model describes the process more accurately than the linear model.

APPLICATION OF NONLINEAR MODEL TO A TIME OPTIMAL CONTROL PROBLEM

Optimal control theory is an area in which an accurate process model is quite important. Accordingly we now apply our nonlinear model to a time optimal control problem involving the reactor previously discussed. The problem is to change the reactor's outlet concentration from its nominal steady state to a new level. The time to do this is to be minimized. Expressing this in mathematical form, it is desired to drive the system from

$C(0) = 0.0$ g-mol/cc
$\dot{C}(0) = 0.0$ g-mol/cc-sec

to

$C(TF) = 0.249 \times 10^{-3}$ g-mol/cc
$\dot{C}(TF) = 0.0$ g-mol/cc-sec

while minimizing

$$J = \int_0^{TF} dt$$

where: TF = final time, secs.

The forcing function, manipulated variable, is restricted by

-3.0 cc/sec $\leq M(t) \leq 5.0$ cc/sec

Optimal control theory, based on Pontryagin's minimum principle, concludes that for linear systems the time optimal is bang-bang in nature. That is, the controls will assume only extremal values switching from one extreme to the other. For nonlinear systems of the form

$$\frac{d\bar{x}}{dt} = \bar{f}(\bar{x}) + \bar{F}(\bar{x})\bar{M}(t)$$

where: $\bar{x}$ = state variable vector
$\bar{f}(\bar{x})$ = vector, which is a function of the state variable, may be nonlinear
$\bar{F}(\bar{x})$ = matrix, which is a function of the state variable, may be nonlinear
$\bar{m}(t)$ = forcing function vector

The time optimal control is also bang-bang in nature [6].

For linear systems the maximum number of switches of the controls can be predicted. They are given by:

Switches = n - 1

where: n = order of the system

For nonlinear systems there is no way to predict the maximum number of switches. A more extended review of time optimal control is given by Koppel [6] and Lapidus [7].

In the next two sections the optimal control law M*(t) is calculated for this problem based on the nonlinear and the linear models. These controls were applied to the simulated reactor and the results were compared. The optimal control law, M*(t), consists of a time, TS, for switching between the extreme values of the manipulated variable, and a final time, TF, for returning to conventional control.

Solution of Problem

The nonlinear model of the stirred reactor is of the form:

$$f(C(t))\frac{d^2C(t)}{dt^2} + g(C(t))\frac{dC(t)}{dt} + C(t) = k(C(t))M(t) \quad (13)$$

where: $f(C(t)) = 870.42 + 5383.5 \times 10^3 C(t)$, secs^2
$g(C(t)) = 66.35 + 172.9 \times 10^3 C(t)$, secs
$k(C(t)) = 0.613 \times 10^{-4} + 0.0891 C(t)$, g-mol/cc (cc/sec)

To obtain the control law to accomplish the solution of the problem the minimum principle technique is used [8]. Equation (13) can be reduced to two first order equations:

$$\frac{dC(t)}{dt} = \dot{C}(t) \quad (14)$$

and

$$\frac{d\dot{C}(t)}{dt} = [k(C(t))M(t) - C(t) - g(C(t))\dot{C}(t)]/f(C(t)) \quad (15)$$

Forming the Hamiltonian:

$$H = 1 + \lambda_1(t)\dot{C}(t) + \lambda_2(t)[k(C(t))M(t) - C(t) - g(C(t))\dot{C}(t)]/f(C(t)) \quad (16)$$

from which the adjoint equations can be obtained:

$$\dot{\lambda}_1(t) = -\frac{\partial H}{\partial C(t)}$$

$$= -\lambda_2(t)\frac{f(C(t))[M(t)\frac{\partial k(C(t))}{\partial C(t)} - 1 - \dot{C}(t)\frac{\partial g(C(t))}{\partial C(t)}]}{f(C(t))^2}$$

$$- \frac{[k(C(t))M(t) - C(t) - g(C(t))\dot{C}(t)]\frac{\partial f(C(t))}{\partial C(t)}}{f(C(t))^2} \quad (17)$$

$$\dot{\lambda}_2(t) = -\frac{\partial H}{\partial \dot{C}(t)} = -\lambda_1(t) + \lambda_2(t)g(C(t))/f(C(t)) \quad (18)$$

From the Hamiltonian, equation (16), we notice that to minimize H subject to the restriction on the forcing function, M(t), we need:

$$M^*(t) \begin{array}{l} 5.0 \text{ cc/sec if } \frac{\lambda_2(t)k(C(t))}{f(C(t))} < 0 \\ -3.0 \text{ cc/sec if } \frac{\lambda_2(t)k(C(t))}{f(C(t))} > 0 \end{array} \quad (19)$$

Equation (19) indicates the correct control action. When

$$\frac{\lambda_2(t)k(C(t))}{f(C(t))} = 0$$

there is no indication of the correct action. This situation is called singularity. When this happens the minimum principle does not explicitly define the necessary control [8]. In this problem singularity can happen by having:

$k(C(t)) = 0$
$f(C(t)) = \infty$

or

$\lambda 2(t) = 0$

For the first condition to happen:

$0.613 \times 10^{-4} + 0.0891\, C(t) = 0$

$C(t) = -0.687 \times 10^{-3}$ g-mol/cc

but

$C(t) = c(t) - c_s$

$c(t) = C(t) + c_s$

$ = -0.687 \times 10^{-3} + 0.162 \times 10^{-3}$

$c(t) = -0.525 \times 10^{-3}$ g-mol/cc

which is a physical impossibility. Therefore, the first condition will never happen.

For the second condition to happen:

$66.35 + 172.9 \times 10^{-3}\, C(t) = \infty$

$C(t) = \infty$

and

$c(t) = \infty$

but this condition cannot happen since the largest value of $c(t)$ that can be obtained is the inlet value, no reaction occurring. This value is 0.65×10^{-2} g-mol/cc. Therefore, the second condition will not occur either.

For true optimal control there is one other condition, $H = 0$ at all times.

$$0 = 1 + \lambda 1(t)\dot{C}(t) + \lambda 2(t)[k(C(t))M(t) - C(t) - g(C(t))\dot{C}(t)]/f(C(t)) \quad (20)$$

The following way to solve this problem is suggested by Siebenthal and Aris [9]. At $t = TF$ from equation (20):

$0 = 1 + \lambda 2(TF)[k(C(TF))M(TF) - C(TF)]/f(C(TF))$

$\lambda 2(TF) = \dfrac{-f(C(TF))}{k(C(TF))M(TF) - C(TF)}$

The value of $M(TF)$ is -3.0 cc/sec. Therefore:

$$\lambda 2(TF) = \dfrac{-f(C(TF))}{-3.0 k(C(TF)) - C(TF)} \quad (21)$$

Now we start at TF and integrate equations (14), (15), (17) and (18) backwards. For this we have

$C(TF) = 0.249 \times 10^{-3}$ g-mol/cc

$\dot{C}(TF) = 0.0$ g-mole/cc-sec

$\lambda 2(TF) = -f(C(TF))/(-3.0k(C(TF)) - C(TF))$

$\lambda 1(TF) =$ first trial value

As soon as $\lambda 2(t)$ changes sign we switch from $M(t) = -3.0$ cc/sec to $M(t) = 5.0$ cc/sec. The integration is continued until $C(t) = C(0) = 0.0$ g-mol/cc, the initial value of the state variable. If at this point $\dot{C}(t) = \dot{C}(0) = 0.0$ g-mol/cc-sec then the value used for $\lambda 1(TF)$ is the correct one. If not, a new value of $\lambda 1(TF)$ must be guessed and the calculations repeated. When the correct value of $\lambda 1(TF)$ has been found, the switching time, TS, and the final time, TF, can be calculated from the results of the backward integration. Pattern Search optimization technique [10] was used to obtain the correct value of $\lambda 1(TF)$. The solution to this problem is then [11]:

$0 \leq t < TS$; $M(t) = M1$

where $M1 = 5.0$ cc/sec
$TS = 83.09$ sec

and

$TS \leq t \leq TF$; $M(t) = M2$

where $M2 = -3.00$ cc/sec
$TF = 90.29$ sec

The same procedure was repeated to obtain the optimal control, TS and TF based on a linear model of the reactor. This linear model is given by equation (10). The results are:

$TS = 120$ secs

and

$TF = 122.5$ secs

COMPARISON OF RESULTS

The switching and final times obtained in the previous section were applied to the simulated reactor. Table 2 presents the results. Column 1 gives the model used to obtain TS and TF. Columns 2 and 3 show the corresponding times. Column 4 gives the value of the state variable, reactor's outlet concentration, at the final time, TF, when conventional control is resumed. This is the "actual" value of the state variable, not a perturbed variable. The desired new level is 0.411×10^{-3} g-mol/cc. Column 5 shows the percent error in final value of the state variable. The nonlinear model definitely gives much better results than the linear model. Columns 6 and 7 present an interesting comparison. After time TF the control is returned to a conventional PI, or PID, controller. From here on the controller must regulate the manipulated variable, coolant flow rate, to maintain set point. If the control is given back to the controller without doing anything to the manipulated variable, the controller would have to increase this variable immediately to keep set point. This results in oscillations of the system and therefore in longer settling time. To avoid these oscillations, the value of the manipulated variable needed to maintain set point is normally calculated by the model. Then the manipulated variable is set to this value before control is returned to the controller. The value needed to maintain set point is

M(TF) = 8.13 cc/sec

This is the actual value, not a perturbed value. Column 6 presents the values calculated from the models. Column 7 gives the percent error between these values and the desired value. Again, the nonlinear model gives much better results than the linear one. Figure 9 shows the response of the simulated reactor when the control laws are applied and control is returned to a PI feedback controller after time TF.

SUMMARY AND CONCLUSIONS

A second order nonlinear process model has been developed. The model contains six parameters. These parameters are easily evaluated by performing three or more step tests in the process unit. The model was tested by application to a simulated stirred tank chemical reactor. Comparison of the nonlinear model with a conventional second order linear model of the reactor showed that the non-linear model was superior.

An important point worth mentioning is the following. The basic idea in developing this nonlinear model is to fit K, $\tau_1 \tau_2$ and $\tau_1 + \tau_2$ versus C(t). In the examples shown in this paper, these relationships happen to be linear. This is not necessarily general. That is, these relationships might not always be linear. Still, the basic idea behind this model holds.

NOMENCLATURE

a1,a2,a3	Parameters in the second order nonlinear model
b1,b2,b3	Parameters in the second order nonlinear model
c(t)	State variable
c_{ss}	Nominal steady state of the state variable
C(t)	State variable, perturbed form; C(t) = c(t) - c_{ss}
C_{ss}	Steady state value of perturbed state variable
H	Hamiltonian function
K	Gain, parameter in mathematical model
m(t)	Manipulated variable
m_{ss}	Steady state value of manipulated variable
M(t)	Manipulated variable, perturbed form; M(t) = m(t) - m_{ss}
s	Laplace transform variable
t	Time
TF	Final Time
TS	Switching Time
λ_1, λ_2	Adjoint variables in optimal control problem
τ_1, τ_2	Time constants in second order linear model

REFERENCES

[1] Smith, C. A. and F. R. Groves, Jr., "Empirical Nonlinear Process Models Applied to Chemical Reactor Control", *Instrumentation in the Chemical and Petroleum Industries*, Vol. 9, ISA, 1973.

[2] Smith, C. A. and F. R. Groves, Jr., "Dead Time Compensation Based on Empirical Nonlinear Process Models", International Conference and Exhibit, Houston, Texas. ISA, 1973.

[3] Murrill, P. W., *Automatic Control of Processes*, International Textbook Co., Scranton, Pa., 1967, pp. 296-298.

[4] Koppel, L. B., *Introduction to Control Theory*, Prentice Hall, Inc., Englewood Cliffs, N.J., 1968, pp. 351-353.

[5] Kern, D. Q., *Process Heat Transfer*, McGraw Hill Co., New York, N.Y., 1950, p. 203.

[6] Koppel, L. B., op. cit., pp. 226-228.

[7] Lapidus, L. and R. Luus, *Optimal Control of Engineering Processes*, Blaisdell Publishing Co., Waltham, Mass., 1967, pp. 144-150.

[8] Koppel, L. B., op. ct., pp. 254-258.

[9] Siebenthal, C. D. and R. Aris, "The Application of Pontryagin's Methods to the Control of a Stirred Reactor", *Chemical Engineering Science*, Vol. 19, 1964, pp. 729-746.

[10] Wilde, D. J. and C. S. Beightler, *Foundation of Optimization*, Prentice Hall, Inc., Englewood Cliffs, N.J., 1967, pp. 307-313.

[11] Smith, C. A., "Nonlinear Models of Chemical Processes", Ph.D. Dissertation, Louisiana State University, December 1972, p. 119.

APPENDIX A

SIMULATION OF STIRRED TANK CHEMICAL REACTOR

The stirred tank reactor example was obtained from Koppel [4]. The response of this reactor is quite nonlinear arising mainly from the Arrhenius term in the kinetic expression and the log mean temperature difference term in the heat transfer expression.

As shown in Figure A-1, this reactor is a stirred tank in which an irreversible exothermic reaction of the type

$$A \xrightarrow{k} B$$

with first order kinetics is taking place. This reactor is cooled by means of a cooling coil. The reaction rate constant is:

$$k = k_0 \exp(-E/RT) \qquad (A-1)$$

where: k = Arrhenius reaction rate constant (sec^{-1})
k_0 = frequency factor (7.86×10^{12} sec^{-1})
E = activation energy (2800 cal/g-mole)
R = gas constant (1.987 cal/g-mole-°K)

The mass balance on the reactor contents, assuming uniform mixing is:

$$v \frac{dCA}{dt} = FCA_0 - FCA - VkCA \tag{A-2}$$

where: V = material volume (1000 cc)
CA = outlet concentration of component A (g-mole/cc)
CA_0 = inlet concentration of component A (6.5×10^{-2} g-mole/cc)
C = volumetric flow (cc/sec)

The energy balance on the contents is:

$$v\rho cp \frac{dT}{dt} = F\rho cp (T_0 - T) - \Delta HVkCA - \frac{UA (T_c - T_{co})}{\ln[(T-T_{co})/(T-T_c)]} \tag{A-3}$$

where: ρ = density of contents (1 gm/cc)
cp = heat capacity of contents (1 cal/gm°K)
T = temperature of reactor and exit stream (°K)
T_0 = temperature of inlet stream (350°K)
$-\Delta H$ = exothermic heat of reaction (2700 cal/g-mole)
UA = overall heat transfer coefficient times cooling coil area (cal/sec-°K)
T_{co} = inlet coolant temperature (300°K)
T_c = exit coolant temperature (°K)

It has been assumed in this heat balance that no heat transfer occurs with the surroundings and that the physical properties of the inlet and exit streams are identical.

The lumped energy balance for the cooling coil is:

$$\frac{V_c \rho_c C_{Pc}}{2} \frac{dT_c}{dt} = \frac{UA (T_c - T_{co})}{\ln[(T-T_{co})/(T-T_c)]} - F_c \rho_c C_{Pc} (T_c - T_{co}) \tag{A-4}$$

where: v_c = coil volume (100 cc)
ρ_c = coolant density (1 gm/cc)
c_{Pc} = coolant heat capacity (1 cal/gm-°K)
F_c = coolant flow ($0 \leq F_c \leq 20$ cc/sec)

The coolant flow is used as forcing function and the exit concentration of component A is the state variable of interest.

This reactor was simulated on the digital computer. A fourth order Runge Kutta method was used to integrate Equations A-2, A-3 and A-4.

TABLE 1

$M(T)$, cc/sec	C_{ss}, g-mole/cc	K, g-mole/cc-(cc/sec)	τ_1 secs	τ_2, secs
-2.0	-0.0989×10^{-3}	0.494×10^{-4}	8.26	41.0
1.5	0.1056×10^{-3}	0.704×10^{-4}	22.68	57.2
3.0	0.249×10^{-3}	0.830×10^{-4}	27.70	81.7

TABLE 2
TIME OPTIMAL CONTROL LAW RESULTS

MODEL	TS (SECS)	TF (SECS)	$c(TF)$ (g-mol/cc)	ERROR (%)	$M(TF)$ (cc/sec)	ERROR (%)
LINEAR	120.0	122.5	0.530×10^{-3}	29.0	8.84	8.70
NONLINEAR	83.09	90.29	0.426×10^{-3}	3.64	8.11	0.24

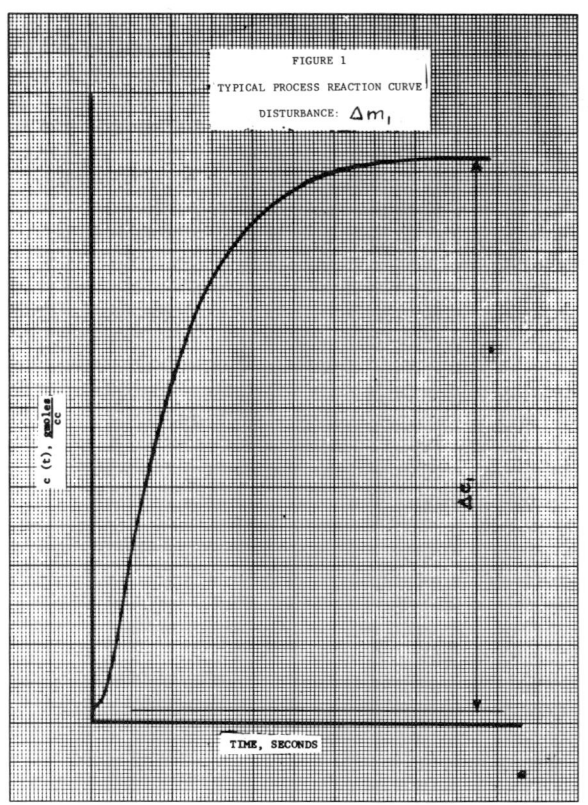

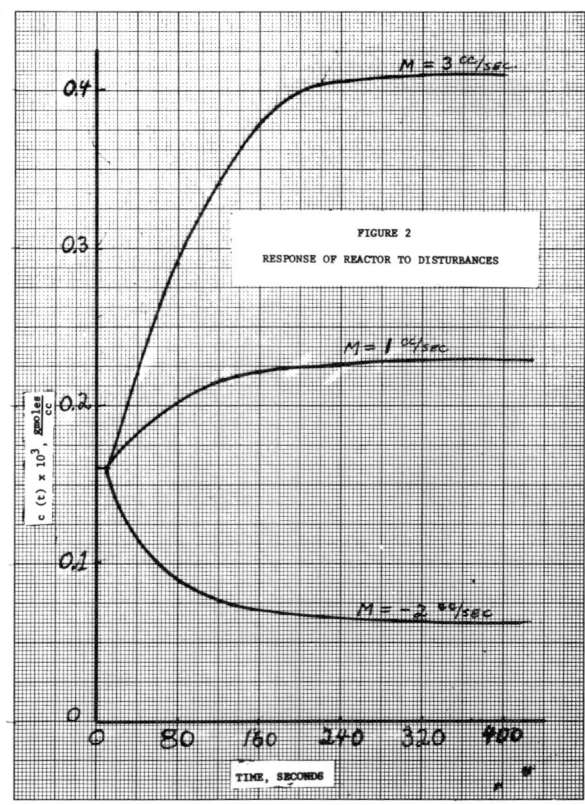

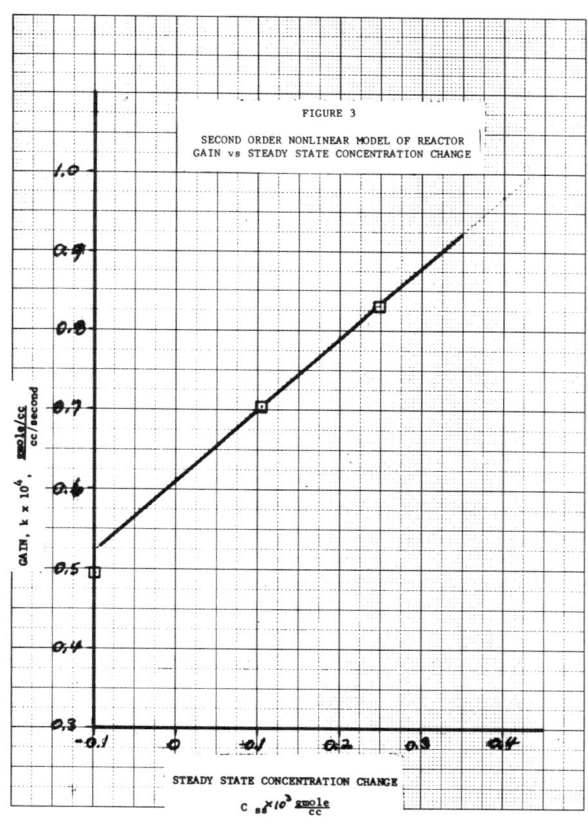

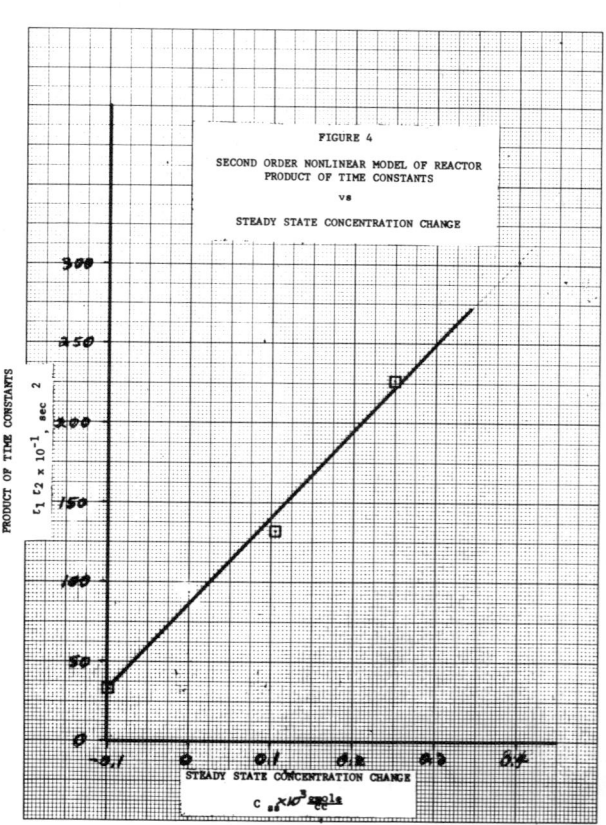

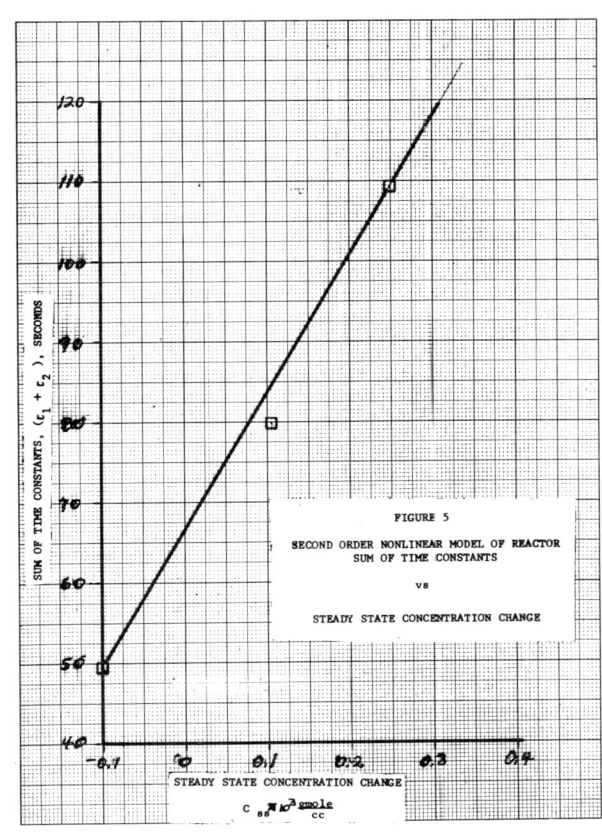

FIGURE 5

SECOND ORDER NONLINEAR MODEL OF REACTOR
SUM OF TIME CONSTANTS
vs
STEADY STATE CONCENTRATION CHANGE

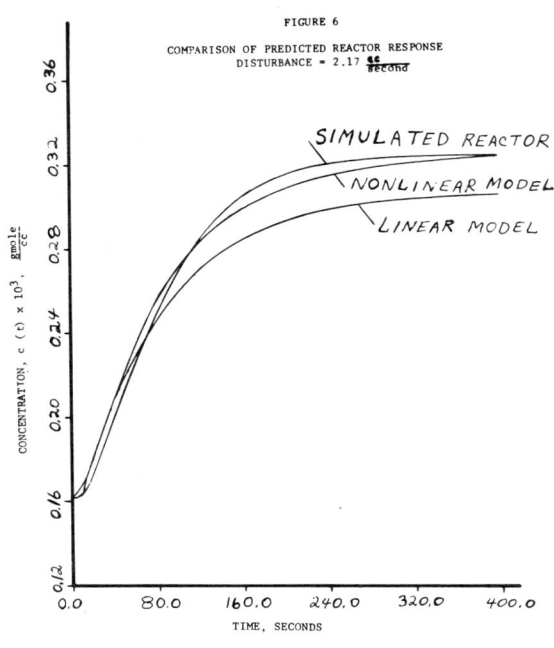

FIGURE 6

COMPARISON OF PREDICTED REACTOR RESPONSE
DISTURBANCE = 2.17 $\frac{cc}{second}$

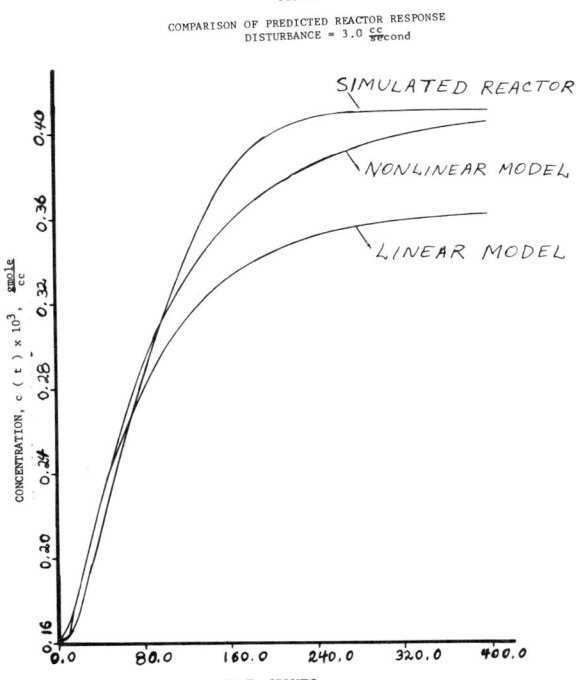

FIGURE 7

COMPARISON OF PREDICTED REACTOR RESPONSE
DISTURBANCE = 3.0 $\frac{cc}{second}$

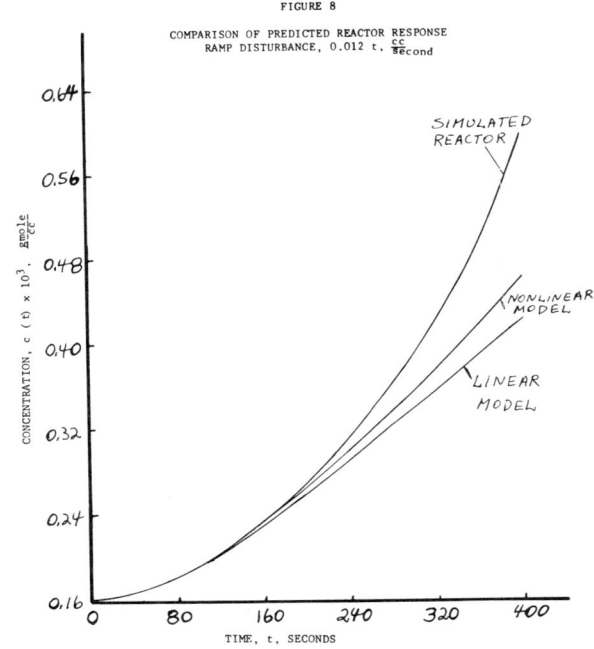

FIGURE 8

COMPARISON OF PREDICTED REACTOR RESPONSE
RAMP DISTURBANCE, 0.012 t, $\frac{cc}{second}$

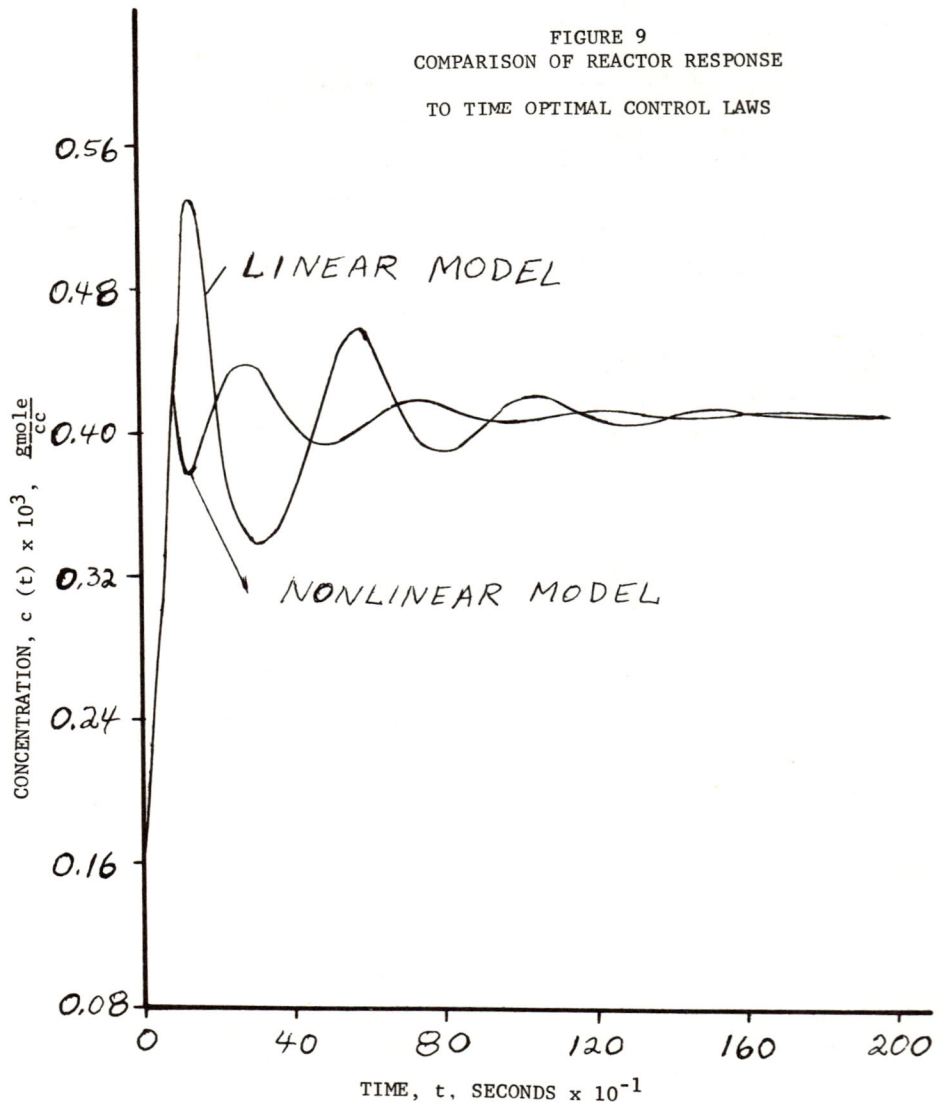

©ISA, 1977
ISBN 87664-363-2

EVALUATION OF VARIOUS CONTROL STRATEGIES
FOR A POWER PLANT FUEL OIL HEATING CONTROL SYSTEM

Tom P. Davis
Duke Power
Charlotte, N.C.

Carlos A. Smith
Chemical Engineering
University of South Florida
Tampa, Florida 33620

ABSTRACT

This paper covers the application of a digital computer simulation of a power plant fuel oil heating system to evaluate various proposed control strategies. Control system response to various process pertubations are presented and general recommendations are made regarding improvements to the existing control system.

BACKGROUND

The fuel oil heating system upon which this work is based is typical of a large modern power plant designed only to handle residual and crude oils. No provisions were made for coal firing. Residual fuel oil is delivered to the plant by pipeline from another power station located approximately 30 miles away, and is stored in two 250,000 barrel capacity storage tanks. From these tanks the oil is pumped through fuel oil heaters and heated to obtain the desired viscosity, then to booster pumps which raise the oil to high pressure before reaching the burner headers at the boiler. The boiler is a tangentially-fired unit with twenty recirculation type burners located at five elevations in the furnace. This drum-type boiler is fired at a full load rate of 4200 million BTU/hr and supplies steam to a nominal 515 MW turbine-generator unit.

Since heavy fuel oils are highly viscous, almost solidified at room temperature, the fuel oil must be heated in fuel oil heaters to obtain the desired viscosity prior to reaching the burners. For fuel oil heating, a fuel oil heater assembly consisting of 80 hairpin concentric-pipe heat exchanger elements stacked vertically 4 units high and 20 units wide is utilized. The heat exchanger elements are connected in 20 parallel passes, each pass consisting of 4 heat exchanger elements in series. Each of the 20 parallel passes are connected together by oil and steam manifolds with shutoff valves for each parallel pass. The fuel oil heater is connected counterflow, with oil entering at the bottom and flowing upward through each parallel pass. Saturated steam entering at the top of the fuel oil heater assembly finally exits as subcooled condensate. Oil flows in the shell side over aluminum fins and is heated by the steam/condensate mixture flowing in the inner pipe. At rated design conditions, No. 6 fuel oil, at a flow rate of 600 GPM flows through the shell side and is heated from 120°F to 240°F. Saturated steam entering at 366°F and 150 psig flowing at a mass flow rate of 22,000 lb/hr is condensed and leaves the heater at 210°F. Total design heat transfer at rated conditions is 19,000,000 BTU/hr. The outer shell of each heat exchanger element consists of 3 1/2 inch schedule 40 steel pipe and the inner pipe consists of 1 1/2 inch schedule 40 steel pipe. Bonded to the inner pipe are 16 aluminum fins, longitudinal and continuous over the entire length of the heat exchanger. The hairpin elements are approximately 22 feet long, providing 176 total feet of pipe per parallel path of 20 total parallel paths.

Since the heat transfer and oil flow are relatively high in this large installation, steam flow to the heater cannot be controlled due to the risk of thermal shock under transient conditions. Instead, the rate of heat transfer is varied by varying the heat transfer surface area available to make contact with the incoming steam. This is done by controlling the condensate level in the heater by means of a condensate drain control valve. This method of control is much slower than controlling steam flow, but eliminates the concern regarding thermal shock.

Oil leaving the fuel oil heater immediately enters the fuel oil booster pumps which raise the oil to high pressure before traveling through the fuel oil supply piping to the main fuel oil heaters at the boiler.

INITIAL DESIGN

For an initial design, a simple single loop feedback control system was proposed, as shown in Figure 1. Since excessive piping deadtime and thermal lag was to be expected with the massive fuel oil heating system, and since considerable heat is added to the oil by the fuel oil booster pumps at low loads on the plant, the decision was made to locate the viscometer as close to the burner heaters as possible to maximize accuracy.

A continuous electrical signal, proportional to viscosity, supplied by a transmitter located at the installation was compared with a setpoint signal to develop an error signal based upon the desired viscosity. The error signal was then used to control the position of the fuel oil heater condensate drain valve. An increase in the viscosity signal above setpoint indicates that the fuel oil temperature is too low and the drain valve is opened further to increase heat transfer surface in the heater. The opposite response occurs for a low viscosity signal.

For a review of the available industrial viscometers, the reader is referred to the literature (1,2,4).

With the control problems of deadtime and thermal lag inherent within the fuel oil heating system, a means of examining and evaluating various control schemes was required. A computer simulation of the fuel oil heating control system was used as an effective tool to attack this engineering problem. (3,5)

The purpose of the particular computer simulation was to develop a simple but sufficiently accurate model of the fuel oil heating system to allow careful engineering evaluation of the system and alternative control systems. The philosophy of the computer simulation was not to develop a highly accurate, rigorous mathematical analysis in microscopic detail, but rather a workable but sufficiently accurate approach to a practical engineering problem, keeping in mind the model limitations.

In most computer simulations of dynamic systems, a mathematical model must first be obtained that defines the input-output relationships of a process as functions of time. Usually these mathematical relationships are in the form of simultaneous differential equations. These differential equations are determined either empirically, from the operating system or a similar operating system, or analytically, from the physical laws governing the heat transfer and fluid flow dynamics of the system. The computer simulation described in this work was derived from a combination of these two techniques.

With the computer simulation, not only could existing problems be analyzed, but proposed alternative control schemes could be compared on a performance basis. Studies were made regarding fuel oil heating system control problems and several alternative control schemes were proposed to compensate for these problems. With the computer simulation, these alternative schemes were compared on the basis of response to transients and controllability. Sufficient study could be made on each proposed system to make recommendations regarding improvements to the existing control system without trial and error field changes which are costly and time consuming.

CONTROL SYSTEM ANALYSIS AND EVALUATION

To determine the best approach to study the control system and its operating characteristics. The physical parameters which affect the output fuel oil viscosity are listed below:

1. Inlet Steam Conditions
2. Inlet Oil Temperature
3. Viscosity-Temperature Characteristics of Fuel Oil
4. Oil Flow Rate

Plant operating data indicates that Parameter 1., Inlet Steam Conditions, is not likely to change under normal operating conditions, and any changes that occur at all would occur gradually, over a long time period. Since the computer simulation indicated that the fuel oil heating system was not sensitive to minor variations in steam conditions, it was assumed that this parameter was not a factor to consider in fuel oil heating system transients. The same was found to be true with Parameter 2., Inlet Oil Temperature, since the mass of the oil in the 250,000 barrel storage tanks is so large, and causes the mass of oil and its associated heat capacity to behave as a huge thermal "flywheel." Although Parameter 3., Viscosity-Temperature Characteristics of Fuel Oil, is expected to change constantly, changes in the viscosity characteristics in a given shipment of supply oil usually occur over periods of hours and days and not usually on a per-minute basis. It was assumed, for purposes of examination and evaluation of control loop performance, that the fuel oil was of constant consistency. However, Parameter 4., Oil Flow Rate, which has a tremendous effect on fuel oil velocity, is a sensitive parameter in fuel oil heater control. In response to firing rate changes associated with changes in turbine-generator load, fuel oil supply flow swings drastically, at the command of the boiler controls. It has been seen from the computer simulation and from the operating system that fuel oil flow changes play the largest role in the initiation of fuel oil heating system transients.

For these reasons, the existing control system as well as the alternative control schemes were evaluated and compared on the basis of response to oil flow changes and control loop setpoint changes. Since the time constant of the turbine-generator, boiler and fuel oil supply system is so large, and instantaneous step changes in oil flow are impossible, oil flow changes in power plant fuel oil systems resemble ramp changes. For this reason, oil flow changes were incorporated into the computer simulation as ramps of both positive and negative slopes, varying in magnitude. A thorough study was performed at representative low load, average load, and at full load oil flow rates with ramp changes of both low and high magnitude corresponding to small and large load changes.

The response of the existing single loop control system to oil flow ramps was found to be sluggish and difficult to control as shown in Figure 2. In this figure, note that the response improves with higher oil flow as well as with smaller changes in oil flow, as is to be expected. Response of the computer simulation to step changes in setpoint showed similar control difficulties, as can be seen in Figure 3.

It was obvious, from the computer studies of the single loop system, that a faster and more controllable system was needed. For this reason, all of the alternative control strategies were directed toward compensating for fuel oil flow changes and piping deadtime. The following three control schemes were developed and analyzed.

Alternative Scheme No. 1

To compensate for oil flow changes, a modification to the single loop control system was proposed using a feedforward technique. As shown in Figure 4, the drain valve position is compensated during fuel flow transients in an attempt to reduce control loop error. The response of this system to various oil flow ramp changes are shown in Figure 5.

Alternative Scheme No. 2

To compensate for piping deadtime, a dual-loop control alternative was proposed as shown in Figure 6. This alternative involves the installation of a two-inch bypass line and control valve to divert a portion of the fuel oil supply around the heater. The mixing of the cooler bypass oil and the heated oil enables more accurate and much faster fuel oil supply temperature control. By incorporation of the two control loops, one control loop for a viscosity-controlled bypass and the other for temperature control of the heater outlet temperature, the control error due to deadtime is reduced.

The viscosity-controlled bypass allowed a faster and more accurate "trimming control" in addition to the temperature-controlled heater. Temperature control at the heater discharge minimized fuel oil piping deadtime and allowed faster control during oil flow transients. The viscosity-controlled bypass allowed for more accurate control and faster recovery time from heater transients.

The response of this system to various oil flow ramp changes is shown in Figure 7. Likewise, the control system response to step changes in viscosity setpoint is shown in Figure 8.

Alternative Scheme No. 3

To compensate for piping deadtime, as well as oil flow transients, a modification including both Alternative Schemes No. 1 and No. 2 was proposed. This control scheme utilized both fuel oil flow feedforward control as well as dual loop control as shown in Figure 9. The response of this control system to various oil flow changes is shown in Figure 10.

Comparision of Control Systems

Although Alternative Scheme No. 1 exhibits somewhat better control than the existing single loop control system, improvement in control is not dramatic. It was seen that Alternative Scheme No. 1 was the worst of the three alternative control strategies.

However, the dual loop alternatives, schemes No. 2 and No. 3, exhibit far better transient response and shorter settling time than the single loop schemes. At low and intermediate oil flows, Alternative No. 3 shows little if any advantage over Alternative No. 2. At higher oil flows, Alternative No. 3 shows some advantage over Alternative No. 2, but, as in the case of Alternative No. 1 versus the existing single loop system, the improvement in control is insignificant.

RECOMMENDATIONS

It was seen from the computer simulation results that the addition of dual loop control using a controlled bypass around the fuel oil heater provides a definite improvement to the fuel oil viscosity control system. At low loads, the piping deadtime becomes excessive, often as much as ten minutes. Dual loop control provides an inexpensive yet effective improvement to the existing control system to compensate for this deadtime. The addition of oil flow feedforward control could be justified only if the unit were to experience wide load swings with corresponding fuel oil flow transients. Since the unit is a newer, more efficient unit, the operation is typical of a base-loaded unit.

This unit, under normal operation, would pick up load early in the morning and probably would not reduce load until later in the evening. During the day, load changes would normally be only gradual. For this reason, the fuel oil feedforward addition is not necessary. As the unit ages, and as plant operation changes over the years to that of a backup unit, the need for fuel flow feedforward compensation may be realized. At that time, the little required instrumentation and control hardware could be easily added to the fuel oil heating control system.

For immediate improvement to the existing control system, however, the addition of dual loop control using a viscosity-controlled bypass line and a temperature-controlled heater is advised. This control system not only adds to the controllability of the system, but also adds to the reliability. An independent temperature control system controls

the fuel oil heater outlet temperature, so that
any instrumentation failure in the viscosity
control loop, such as clogging or malfunction
of the viscometer, would not cause catastrophic
changes in the fuel oil supply temperature at
the burners. A viscometer, by the nature of
its installation and mechanical complexity, is
more subject to malfunction than a temperature
element. Since the instrument is more subject
to failure, the use of the viscometer in a
"trimming" control loop is a desirable feature.

Modifications to the fuel oil heating system
and the addition of control hardware for dual
loop control would be minimal. It is suggested
by the authors that a design incorporating
a bypass line be added using an equal percentage
control valve with an electro-pneumatic converter
to accept an electrical signal from the viscosity
controller. A local pneumatic temperature control
station could be installed in the vicinity
of the fuel oil heater to control the drain
valve locally. Simple gain control without
integral or derivitive compensation was found
to give the best response for the wide variety
of conditions to be expected.

CONCLUSION

The immediate purpose of the alternative control
scheme evaluation was to develop an improvement
to the existing control system by conventional
means using analog control hardware and incorporating minimal piping changes to the fuel
oil heating system. To state that the solution
presented in this work is a unique optimal
control solution would be a dangerous statement.

There are many analog control and digital computer techniques which may be implemented
to solve difficult control problems. In addition, modification of the process itself to
gain better controllability using conventional
hardware is a possibility that should not be
overlooked.

REFERENCES

1. Davis, T. P., Good, E. M., Smith, C. A., "Fuel Viscosity Control", Instrumentation Technology, Journal of the Instrument Society of America, August, 1975, pp. 41-44.

2. Davis, T. P., Good, E. M., Smith, C. A., "Evaluation of Process Viscometers for Fuel Oil Viscosity Measurement and Control", Proceedings of the Eighteenth Instrument Society of America Power Instrumentation Symposium, May 19-21, 1975, Houston, Texas.

3. Davis, T. P., Good, E. M., Smith, C. A., "Design and Computer Simulation of a Viscosity-Controlled Fuel Oil Heating System", Proceedings of the Eighteenth Instrument Society of America Power Instrumentation Symposium, May 19-21, 1975, Houston, Texas.

4. Davis, T. P., Smith, C. A., "Viscosity Control Improves Heavy Oil Burning", Electric Light and Power, Volume 53, Number 6, June 23, 1975, pp. 56-57.

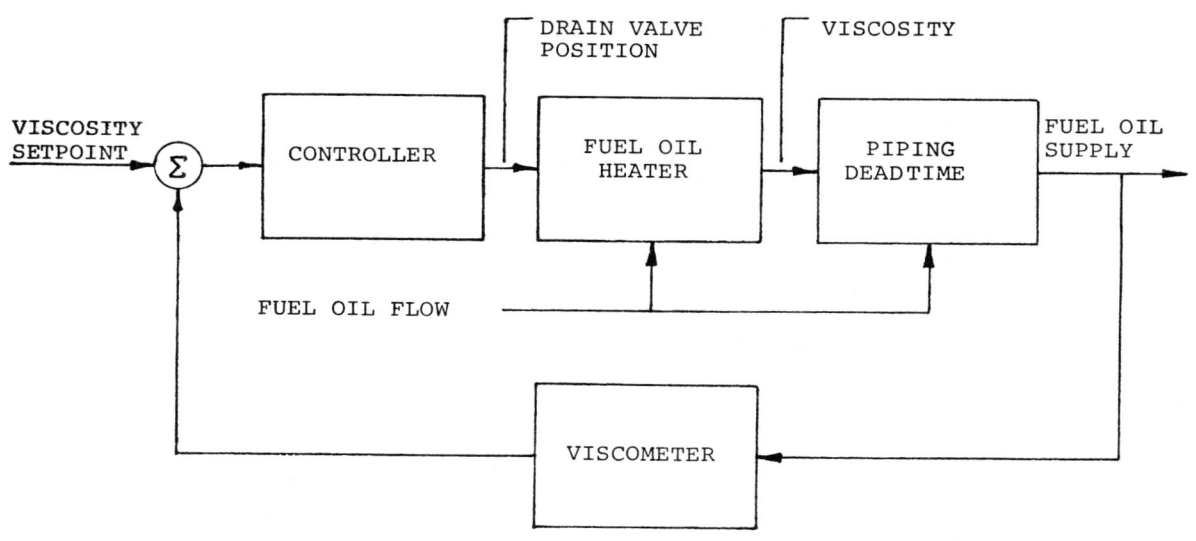

Figure 1

CONTROL SYSTEM BLOCK DIAGRAM

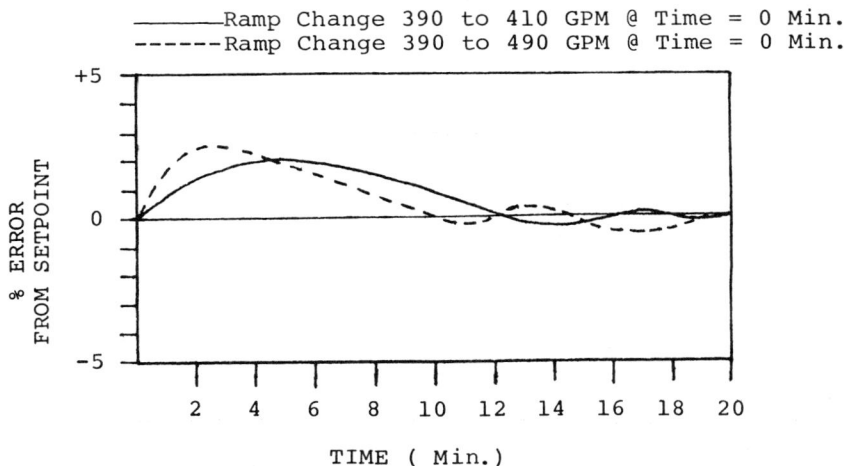

Figure 2

RESPONSE TO OIL FLOW CHANGES-
EXISTING SINGLE LOOP CONTROL SYSTEM

10% STEP INCREASE IN SETPOINT

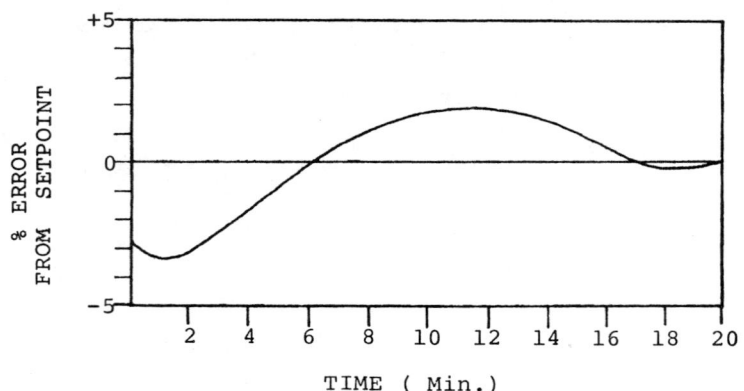

Figure 3

RESPONSE TO VISCOSITY SETPOINT CHANGES
EXISTING SINGLE LOOP CONTROL SYSTEM

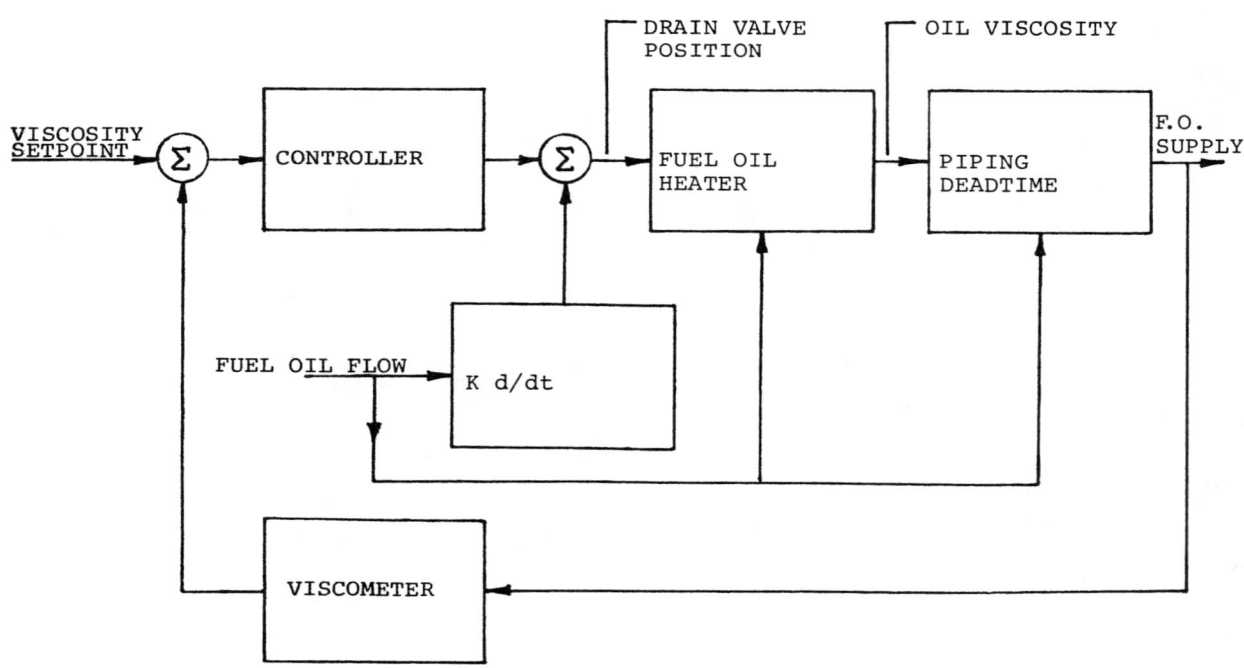

Figure 4

BLOCK DIAGRAM - ALTERNATIVE SCHEME NO.1

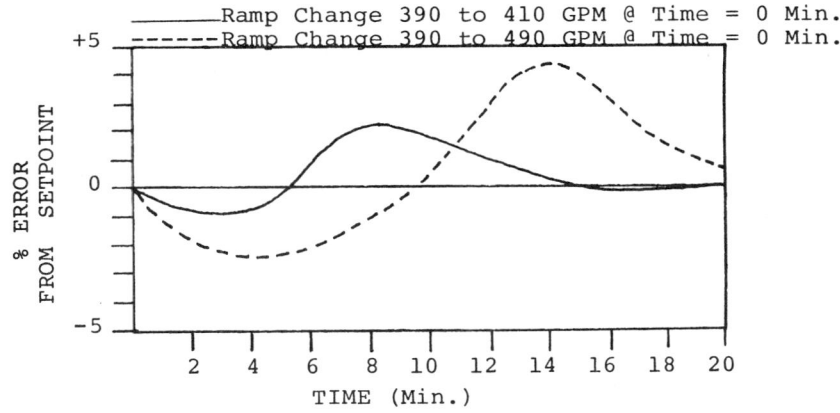

Figure 5

RESPONSE TO OIL FLOW CHANGES-
ALTERNATIVE SCHEME NO. 1

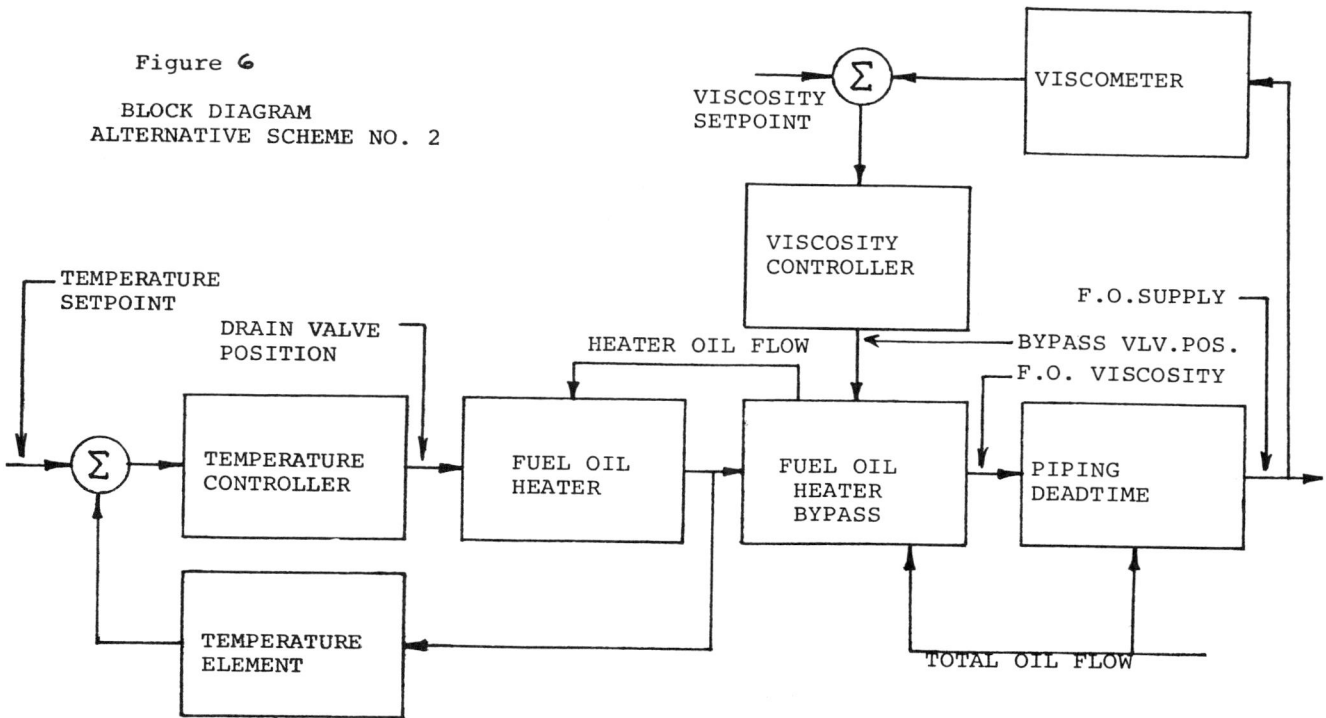

Figure 6

BLOCK DIAGRAM
ALTERNATIVE SCHEME NO. 2

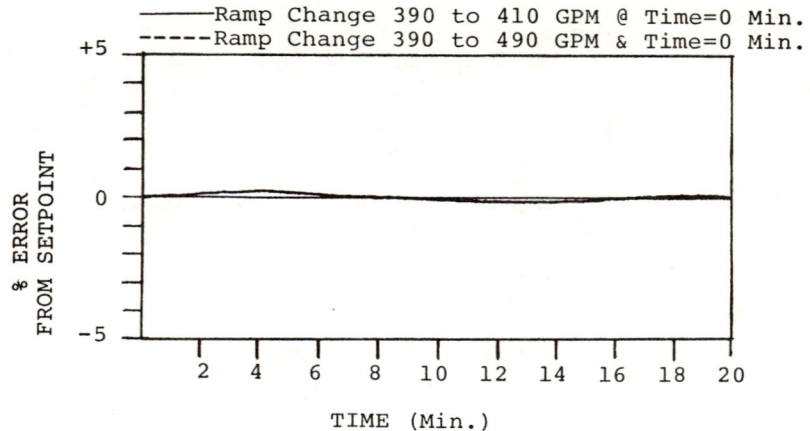

Figure 7

RESPONSE TO OIL FLOW CHANGES
ALTERNATIVE SCHEME NO. 2

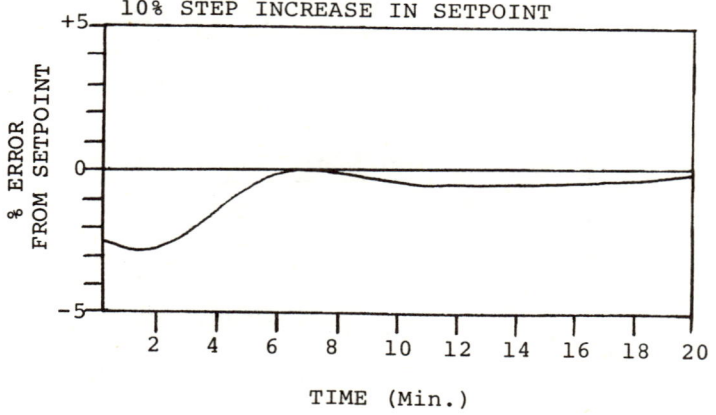

Figure 8

RESPONSE TO VISCOSITY SETPOINT CHANGES-
ALTERNATIVE SCHEME NO. 2

Figure 9

BLOCK DIAGRAM-
ALTERNATIVE SCHEME NO. 3

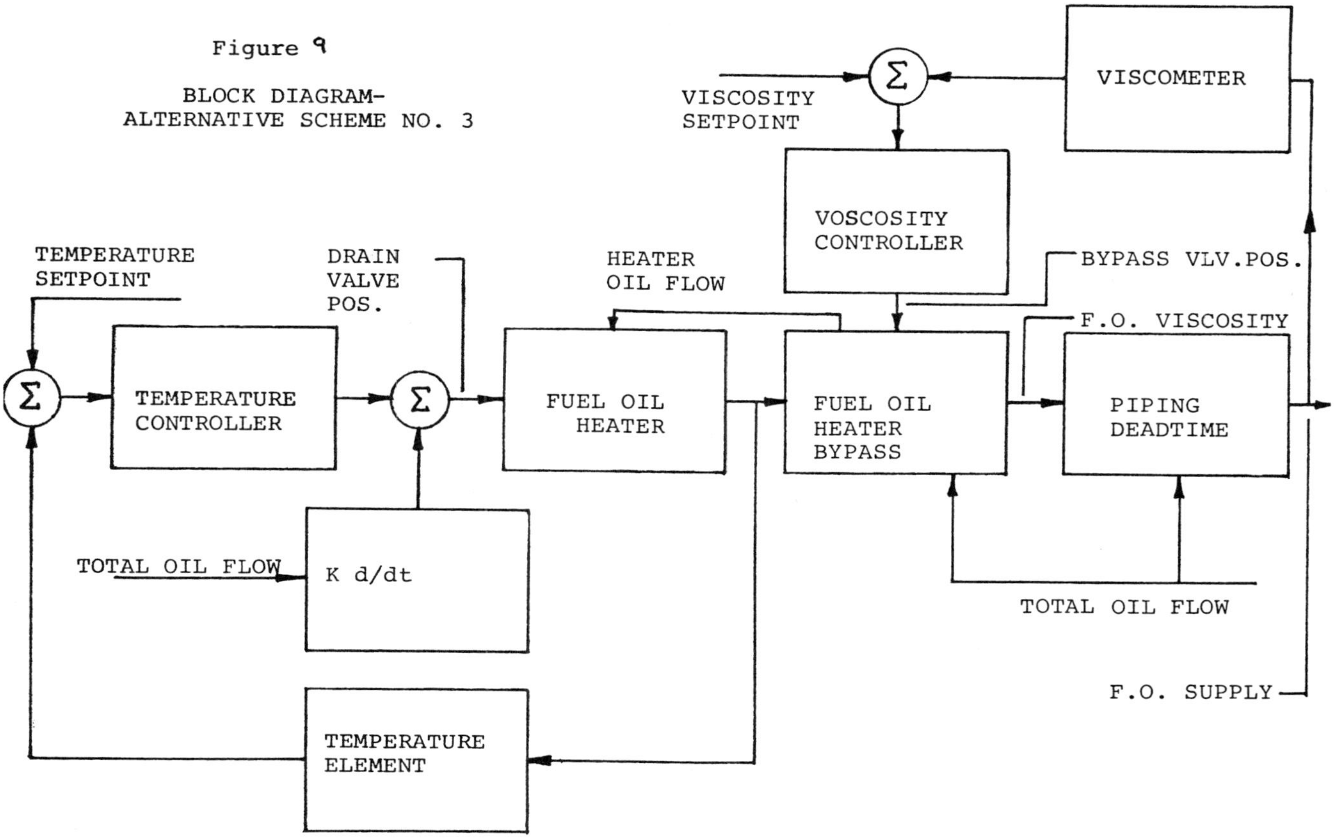

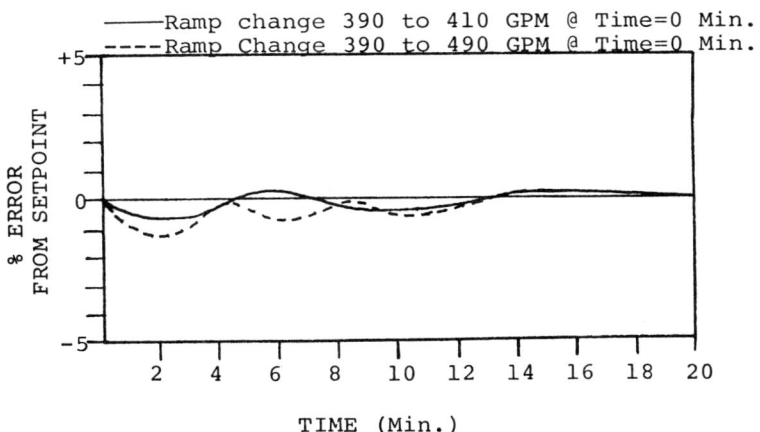

Figure 10

RESPONSE TO OIL FLOW CHANGES-
ALTERNATIVE SCHEME NO. 3

©ISA, 1977
ISBN 87664-363-2

ADAPTIVE CONTROL THROUGH INSTRUMENTAL-VARIABLE
ESTIMATION OF DISCRETE MODEL PARAMETERS

A. T. Touchstone
Dow-Badische Company
Freeport, Texas

Armando B. Corripio
Department of Chemical Engineering
Louisiana State University
Baton Rouge, Louisiana   70803

ABSTRACT

The instrumental-variable (IV) regression technique is used to estimate the parameters of a second-order plus dead-time (transportation lag or time delay) discrete model of the plant. The recursive IV algorithm is shown to behave in a stable drift-free manner in the presence of noise and in a closed-loop arrangement, thus becoming a prime candidate for application to adaptive control.

The paper presents the results of combining IV identification with a Dahlin digital control algorithm for adaptive control of a second-order plant. The scheme is also applied to the temperature control of a continuous stirred tank reactor, a very nonlinear loop. The effect of unmeasured disturbances on the algorithm performance is discussed.

INTRODUCTION

The problem of adaptive control is approached here from the point of view of estimating the parameters of a linear discrete model of a nonlinear process in order to adapt the parameters of a linear discrete feedback controller to the changing dynamic characteristics of the process. The feedback controller parameters are computed as a function of the parameters of the discrete process model. The discrete nature of the model and the controller, as well as the parameter estimation equations, presume that the technique would be implemented on a digital process control computer.

The process model attempts to correlate the dynamic relationship between the process output (measured variable) and its input (controller output signal). In a previous article[1], these authors presented the use of Instrumental-Variable regression to estimate the model parameters in an open-loop configuration: the process input is arbitrarily varied in order to produce the necessary information for parameter estimation. For the estimated parameter values to be used in adaptive control, the identification must be carried out with the loop closed: the process input is manipulated by the feedback controller. In this situation, the process model equation is embedded in a system of two simultaneous equations, of which the other is the control relationship. This paper presents the process control application of a development presented by Klein[2] on simultaneous-equation systems in the study of economic relationships.

THE ADAPTIVE FEEDBACK CONTROL PROBLEM

A sketch of the adaptive feedback control loop is given in Figure 1. The hold device is assumed to be a zero-order hold. The process is usually nonlinear and subject to disturbance and measurement noise. Because of the nonlinear nature of the process, the linear controller must be adapted to the changing dynamic characteristics of the loop if control performance is to be uniform with varying levels of plant operation. The estimator block estimates the model parameters from the sampled values of the process input, u_t, and output, y_t. In the adaptive scheme, the parameter estimates are used to compute the controller parameters.

Process Model

In general, the discrete process model takes the following form:

$$y_t = a_1 y_{t-1} + \ldots + a_p y_{t-p} + b_1 u_{t-M-1} + \ldots + b_q u_{t-M-q} + w_t \qquad (1)$$

The term M in these equations represents the dead-time (transportation lag or time delay) of the system. The term w_t is intended to include all disturbance and measurement noise effects. Because of the presence of past values of the measured variable, y, in the vector x, the term w_t would be correlated noise even if the disturbances are uncorrelated noise. In most process control applications, model order need not be greater than second, i.e., p=q=2.

Feedback Controller

The discrete feedback control algorithm can be represented in practice by the following equation

$$u_t = g_0 e_t + g_1 e_{t-1} + g_2 e_{t-2} + h_1 u_{t-1} + h_2 u_{t-2} \qquad (2)$$

where $e_t = r_t - y_t$.

The algorithm represented by equation 2 can represent a proportional-integral-derivative (PID) control algorithm, given the following substitutions:

$$g_o = K_c(1 + \frac{T}{T_i} + \frac{T_d}{T})$$

$$g_1 = -K_c(1 + \frac{2T_d}{T})$$

$$g_2 = K_c \frac{T_d}{T}$$

$$h_1 = 1$$

$$h_2 = 0 \quad (3)$$

Adaptive Control

In the adaptive control scheme, the parameters of equation 2 can be calculated directly from the parameters of the second-order model (equation 1) by the controller synthesis method proposed by Dahlin[3] or some similar method. In this work, the Dahlin method was selected. For the case of a second-order model with zero dead-time, the controller parameters are obtained by the following formulas:

$$g_o = \frac{1}{b_1} Q$$

$$g_1 = -\frac{a_1}{b_1} Q$$

$$g_2 = -\frac{a_2}{b_1} Q$$

$$h_1 = \frac{b_1-b_2}{b_1}$$

$$h_2 = \frac{b_2}{b_1} \quad (4)$$

where $Q = 1-e^{-T/\beta}$

and β is the time constant of the desired closed-loop response. The constant β is effectively a tuning parameter and is selected by the designer.

PARAMETER ESTIMATION

A detailed discussion of parameter estimation by Instrumental-Variable (IV) regression has been presented by these authors[1]. This method is chosen to estimate the parameters of Equation 1 because it maintains the simplicity of least squares regression while avoiding estimation bias due to measurement noise.

Equation 1 can be written in the following vector form:

$$y_t = \phi x_{t-1} + w_t \quad (5)$$

where

$$\phi = [a_1 \cdots a_p b_1 \cdots b_q]$$

$$x_{t-1}^T = [y_{t-1} \cdots y_{t-p} u_{t-M-1} \cdots u_{t-M-q}]$$

Because the measured output of the plant, y, is corrupted by measurement noise, the presence of past values of the output in the vector x of exogenous variables causes bias on the least-squares estimates of the model parameters. To reduce this bias, the measured output is replaced by an instrumental variable which is highly correlated with the plant output, but uncorrelated with the noise. The output of the model, ξ, was selected in this work for the instrumental variable:

$$\xi_t = \hat{\phi} z_{t-1} \quad (6)$$

where

$$z_{t-1}^T = [\xi_{t-1} \cdots \xi_{t-p} u_{t-M-1} \cdots u_{t-M-q}]$$

is the vector of instruments.

The vector of parameter estimates, $\hat{\phi}$, is then computed by the following recursive formulas, originally proposed by Lee[4]:

$$\hat{\phi}_{t+1} = \hat{\phi}_t + (y_{t+1} - \hat{\phi}_t x_t)[1 + z_t^T P_{t/t-1} x_t]^{-1} z_t^T P_{t/t-1} \quad (7)$$

$$P_{t/t-1} = P_{t-1} + D \quad (8)$$

$$P_t = P_{t/t-1} - P_{t/t-1} x_t [1 + z_t^T P_{t/t-1} x_t]^{-1} z_t^T P_{t/t-1} \quad (9)$$

where $P_{t/t-1}$ is the projection of the weighting matrix P at time t, based on observations up to and including y_{t-1}, and D is a lower bound on P. Matrix D controls the parameter tracking capability of the IV estimator. If D=0, matrix P decreases with time until, eventually, new samples have little effect on the parameter estimates. Each nonzero term of D sets a lower bound on the corresponding term of P, allowing the algorithm to track parameter variations. Matrix D is usually chosen to be diagonal, with each diagonal term corresponding to a different model parameter.

Set-Point Forcing

In order to be used in adaptive industrial control, an identification algorithm must remain stable during periods when the loop is not dynamically forced. However, a forcing function must be applied to the loop in order to produce the necessary information for the estimation of the parameters. It is evident from Figure 1 that the forcing function must be applied in the form of set-point changes, r_t. As shown in this figure, the identification algorithm has as inputs the measured plant output, y_t, and its input, u_t. Now, these variables are not only related through the process side, but also through the feedback controller. By applying the forcing function at the set-point, the plant input variations cannot be completely explained or related by the output variations through the controller, and identification of the plant model parameters results.

If, on the other hand, the dynamic information is produced exclusively by nonrandom unmeasured disturbances, the output variables cannot be completely related to input variations through the plant. In the absence of set-point changes, the identification algorithm would tend to identify the inverse of the controller transfer function.

Changes in set-point are also required to break the possible multicollinearity that would result if a proportional integral (PI) control algorithm is used with a second-order model or a proportional controller is used with a first-order model. In these cases, the input components of vector x become linear combinations of the output components and the system becomes underdetermined in terms of the a and b model parameters.

RESULTS OF CLOSED-LOOP ESTIMATION

In order to quantitatively evaluate the performance of the IV estimator, a second-order linear differential equation was used to represent the plant:

$$2 \frac{d^2c}{dt^2} + 3 \frac{dc}{dt} + c = u(t) \quad (10)$$

The parameters of the second-order discrete model, for T = 1, are given by:

$a_1 = 0.97441$ $b_1 = 0.15482$

$a_2 = -0.22313$ $b_2 = 0.09390$

Since these are the components of the true system parameter vector, ϕ, it is possible to calculate the "estimation error fraction", defined as:

$$\varepsilon = \frac{||\phi - \hat{\phi}_t||}{||\phi - \hat{\phi}_o||} \quad (11)$$

where $\hat{\phi}_t$ and $\hat{\phi}_o$ are the estimates at time t and 0, respectively. The scalar ε is the normalized vector norm of the estimation error in the parameter space. The initial parameter estimates, $\hat{\phi}_o$, are taken as zero. The initial value of the weighting matrix, P_o, is taken as αI, where I is the identity matrix of proper order and $\alpha = 1000$. This causes the IV algorithm to give little weight to the initial parameter estimates.

PID Controller

A digital PID algorithm was first selected for the controller. The controller was tuned to give a minimum integral of time multiplied by the absolute value of the error (ITAE) response to a step change in the set-point[3]. The controller parameters are:

$K_c = 1.776$

$T_i = 3.400$

$T_d = 0.3906$

The set-point function consisted of a series of step changes as shown in Figure 2. The set-point switching interval, λ, is given in integral number of sampling intervals. Figure 3 shows the effect of this parameter on the estimation error fraction ε for closed-loop IV identification. The series was run without disturbances. Notice the stair-step character of the curves which is a result of the improvement of the estimates each time a new set-point change is received.

Figure 4 illustrates the effect of measurement noise on the estimation for $\lambda = 10$. The noise supplied is zero-mean, random, and its magnitude is given by the noise-to-signal ratio, δ, defined by:

$$\delta = \frac{\text{r.m.s. value of noise, } w(t)}{\text{r.m.s. value of the signal, } c(t)} \quad (12)$$

The parameter tracking ability of the IV estimator is demonstrated in Figure 5. The algorithm was required to track the parameter changes which occurred when the process gain was doubled on the 500th sample instant. The lower bound matrix D, set to zero for the preceding cases, was set equal to dI, where I is the identity matrix of appropriate order. Two values of the lower bound element, d, are given in the figure. The d = 0 curve represents the finite-time averaging process. Again, the results are excellent and compare with the results of open-loop estimation.

Adaptive Control

To illustrate the adaptive control loop configuration, the second-order example was used for the unknown process and it was identified utilizing the second-order model as before. But in this instance, the advantage of the on-line estimator's ability to provide current updates of the model parameters can be realized by incorporating a control law which uses this information to select its own parameters. The formulas given as Equation set 4 were used to compute the parameters of a Dahlin algorithm with $\beta = 2$. Figure 6 illustrates the dynamic IV estimator performance in the adaptive environment with the Dahlin control algorithm for the case where the process gain was doubled at the 200th sample and the output was subject to measurement noise, $\sigma_y = 0.025$. As the estimator identifies the new gain, the Dahlin algorithm compensates by halving the controller gain. Clearly, Figure 6 is evidence enough that the IV estimator is most appropriate for the adaptive configuration.

APPLICATION TO A NONLINEAR PROCESS

In order to test the performance of the adaptive control scheme in more realistic situation than that of a known linear plant, the technique was applied to the adaptive temperature control of a simulated stirred tank reactor. A sketch of the reactor is shown in Figure 7. The reactor is represented by a set of three nonlinear differential equations. The details of the simulation have been presented in detail by Chiu[5].

Although the reactor is a third-order system, a second-order model is used for identification. The reason for this is that, if a third-order model was utilized, the control algorithm based on it would be of higher order than a PID algorithm. This is contrary to industrial practice.

The plant model relates the reactor temperature to the cooling water rate, by the following equation:

$$y_t = a_1 y_{t-1} + a_2 y_{t-2} + b_1 u_{t-M-1} + b_2 u_{t-M-2} + C \quad (13)$$

where

$$y = T_R - \overline{T}_R$$

$$u = W_c - \overline{W}_c$$

Note that the linearized nature of the model demands that the input and output variables be expressed in terms of deviations from some reference values, denoted by $\overline{T}_R$ and $\overline{W}_c$. The constant parameter C is included to account for nonzero-mean unmeasured disturbances and for shifting of the operating point from the reference values. In general, best results are obtained when C is small.

The dead-time parameter M must be chosen a priori, or different values tried to determine the one that produces the best results. Although the perfectly stirred reactor contains no dead-time, the fact that the model is of lower order than the process allows for some improvement of model fit by introducing a pseudo-dead-time of one sample time, i.e., M = 1.

Since there are no true system parameters to be compared with the parameter estimates, the IV estimator performance was measured by comparing the open-loop response of the model to that of the process for a step change in cooling water rate. However, the nonlinear nature of the process causes the true response to vary with the magnitude and direction of the input change. This is illustrated in Figure 8, which shows the response of the process to a decrease and to an increase in cooling water rate of 135 kg/min. The best a linear model can do is to produce a response which is the average of the two, since it is supposed to be a linearized representation of the process at the operating point.

PID Controller

The model responses shown in Figure 8 are for parameters estimated with the loop closed with a PID controller and the tuning parameters reported by Chiu[6] for this reactor. The effect of the amplitude, R, of the set-point forcing function is shown in the figure. If the amplitude is too small, the linearized gain is estimated lower than the average. This is probably another effect of the nonlinearity of the system. Once the amplitude is of the order of 1°C, the linearized response becomes insensitive to it. For these runs the lower bound matrix D was set to zero.

Adaptive Control

The need to apply a periodic forcing function to the set-point is an undesirable feature of the IV estimator from the point of view of industrial application. However, examination of the results presented in Figure 3 suggest that it would be possible to obtain the parameter estimates by a sequence of two symmetric pulses, one up and one down, after which the set-point is returned to the normal operating point. This is possible because the parameter estimates remain stable during periods when the system is undisturbed. A pertinent question is to study the effect of unmeasured disturbances entering the process during periods when the set-point is held constant.

The performance of the adaptive control scheme with Dahlin's algorithm is shown in Figure 9. The initial parameter values were those estimated from the nonadaptive run shown in Figure 8. The initial weighting matrix, P_o, was set to zero and the lower bound matrix D was chosen as a diagonal matrix with values proportional to the final values of the P matrix for the nonadaptive run. This makes the estimator sensitive to parameter variations, but also sensitive to process disturbances. This is shown in Figure 9: As the two symmetric pulses in set-point are applied, the estimate of parameter b_1 tracks the variations due to nonlinearities. At the 200th sample, the flow rate of reactants drops, causing the estimate of b_1 to drift to a new value. Although undesirable, this effect is not catastrophic and the adapted algorithm remains stable.

SUMMARY AND CONCLUSIONS

The IV estimator has been shown to perform well in an adaptive scheme as long as a forcing function is applied to the control set-point. This forcing function need not be continuous, but the parameter estimates are sensitive to unmeasured nonzero-mean process disturbances when in the parameter tracking mode, i.e., $D \neq 0$. Although the sensitivity to disturbances may prove to be an unsurmountable problem to the application of this technique in adaptive control, IV estimation is, in the experience of these authors, the most effective technique to obtain process dynamic information from on-line data. To this end it offers the following advantages:

1. Avoids the bias due to measurement noise which is characteristic of ordinary least squares.

2. Requires no information on the statistical properties of the noise.

3. Performs well in the closed-loop configuration.

4. Can be tuned to track parameter variations or to average the results of symmetric pulses.

5. Requires a minimum of computer storage, since it is a recursive formula.

In conclusion, the IV estimation algorithm would be a powerful and useful tool to include in any process control program library.

ACKNOWLEDGEMENT

This work was sponsored by the Air Force Office of Scientific Research, Air Force Systems Command, USAF, under contract 74-2580.

NOTATION

a	model output parameters
b	model input parameters
C	model bias parameter
C_A	reactant concentration, kg/m^3
C_{AO}	inlet reactant concentration, kg/m^3
c	true plant output
D	lower bound matrix
e	control algorithm error
F	reactant flow rate, m^3/min
g	input coefficient in control algorithm
h	output coefficient in control algorithm
K_c	controller gain
M	dead-time
P	weighting matrix
p	number of output terms in model equation
Q	Dahlin controller parameter
q	number of input terms in model equation
r	control set-point
R	amplitude of set-point signal
T	sample time, min
T_c	jacket temperature, °C
T_d	controller derivative time, min
T_i	controller integral time, min
T_o	reactant inlet temperature, °C
T_R	reactor temperature, °C
T_w	water inlet temperature, °C
t	time, min; time index
u	input variable
W_c	cooling water rate, kg/min
w	measurement noise
x	vector of endogeneous variables
y	output variable
z	vector of instrumental variables
β	Dahlin closed-loop time constant, min
δ	noise-to-signal ratio
ε	estimation error fraction
λ	period of set-point signal
φ	vector of model parameters
ξ	model input variable

REFERENCES

(1) Touchstone, A. T. and Corripio, A. B., "Process Identification Utilizing a Sequential Instrumental Variable Regression Algorithm", Fourth Milwaukee Symposium on Automatic Computation and Control, Milwaukee, WI, April 22-24, 1976.

(2) Klein, L. R., _A Textbook of Econometrics_, 2nd Ed., Prentice-Hall, Englewood Cliffs, NJ, 1974.

(3) Smith, C. L., _Digital Computer Process Control_, Intext, Scranton, PA, 1972.

(4) Lee, R. C. K., _Optimal Estimation, Identification and Control_, Research Monograph No. 28, MIT Press, Cambridge, MA, 1964.

(5) Chiu, K. C., Corripio, A. B. and Smith, C. L., "Process Models for Controller Tuning", _Inst. & Cont. Syst._, Vol. 45, No. 1, Jan. 1972, pp. 84-88.

(6) Chiu, K. C., Corripio, A. B. and Smith, C. L., "Digital Control Algorithms. Part I - Dahlin Algorithm", _Inst. & Cont. Syst._, Vol. 46, No. 10, Oct. 1973, pp. 57-59.

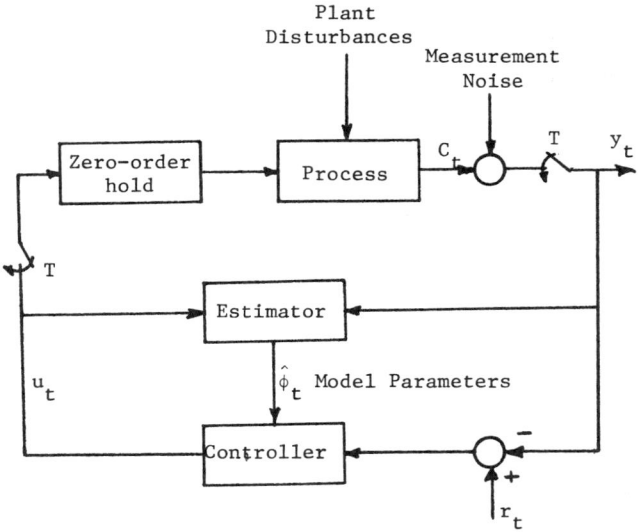

Figure 1. Adaptive Control Configuration

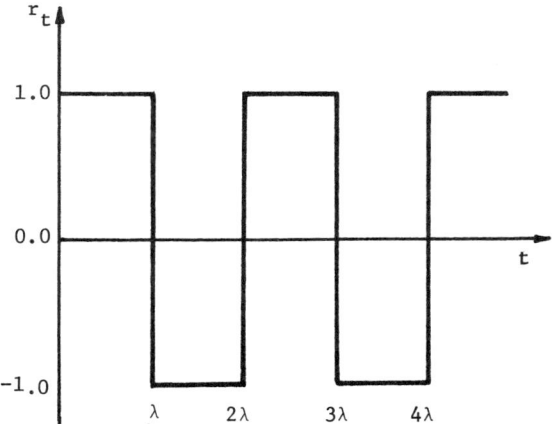

Figure 2: Square-wave set-point function

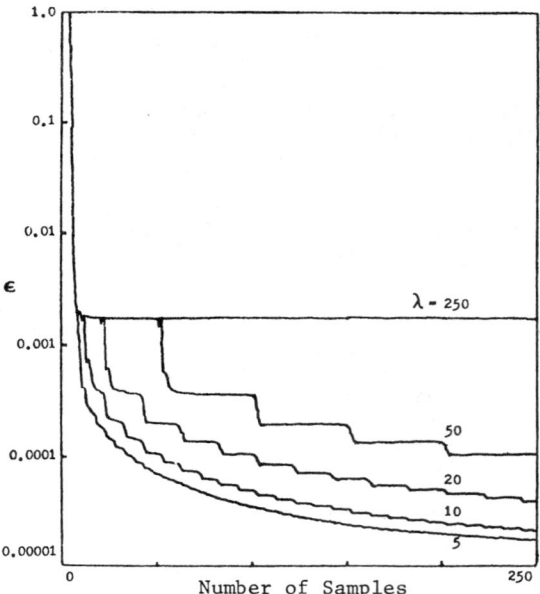

Figure 3. Effect of the set point switching interval on noise-free IV estimation for a PID controller

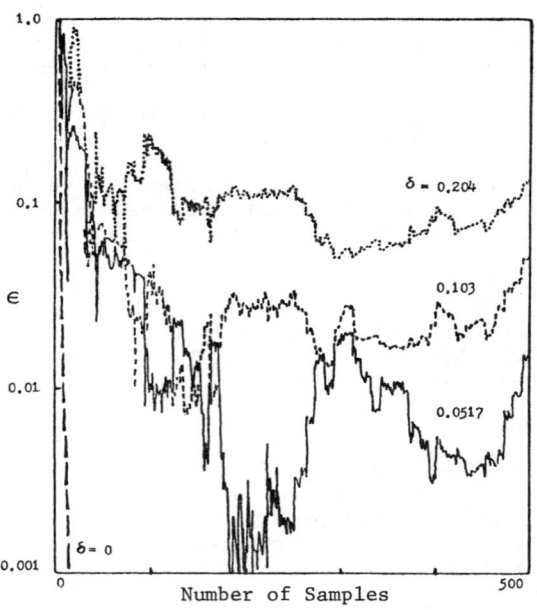

Figure 4. Effect of measurement noise on IV estimation for a PID controller, $\lambda = 10$

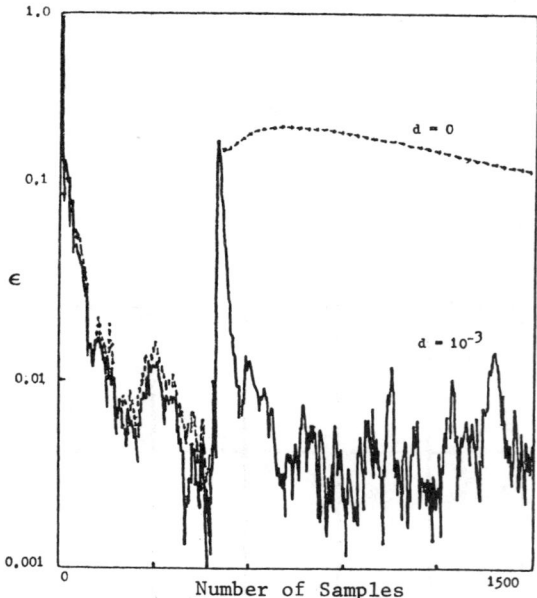

Figure 5. Effect of the variation of the lower bound element d on IV parameter tracking, $\sigma_v = 0.025$

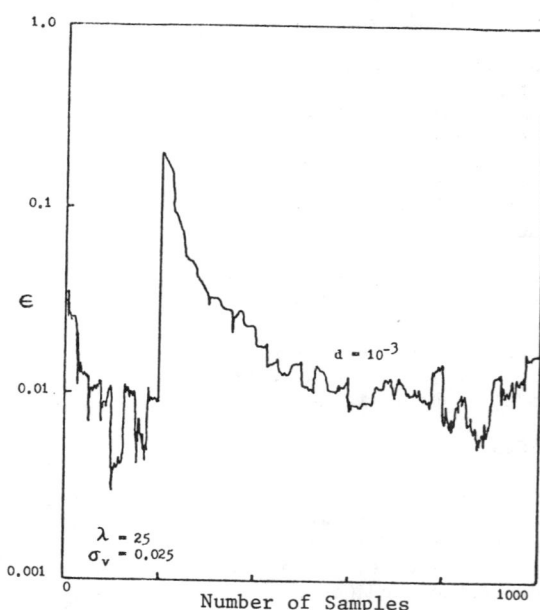

Figure 6. Parameter tracking for adaptive control employing the Dahlin controller. Gain doubled after 200 samples

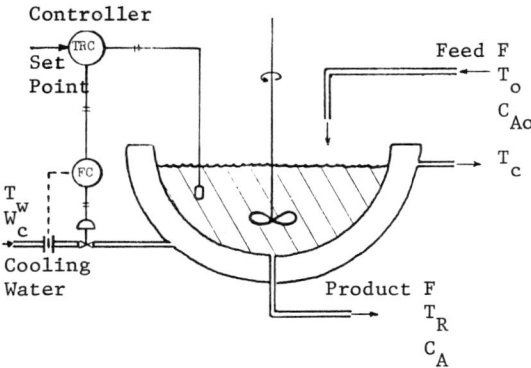

Figure 7. The water-cooled chemical reactor

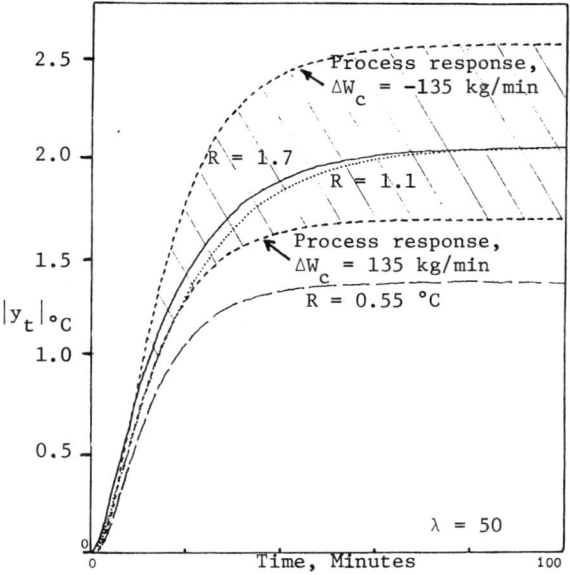

Figure 8. Effect of the set point amplitude on closed-loop IV estimation

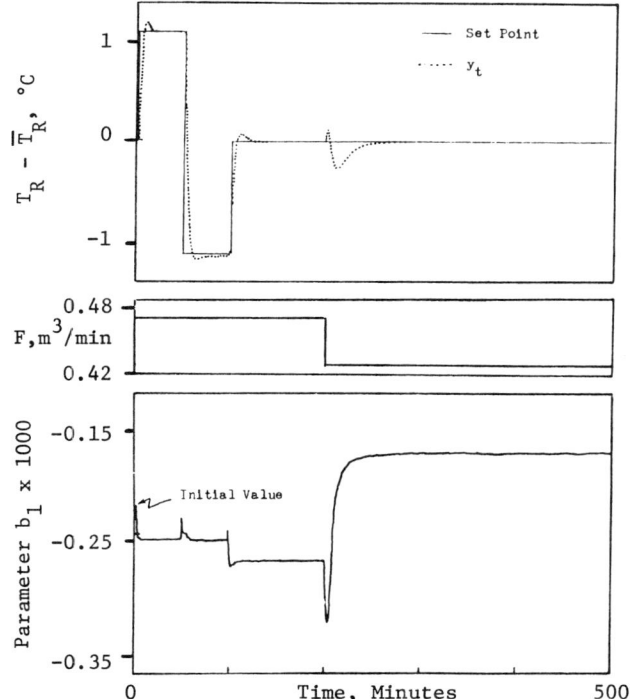

Figure 9. Times series showing the effect of an aperiodic set point function and a step disturbance on the dynamic IV estimation of parameter b_1 for the adaptive configuration

DEVELOPMENT OF MULTIVARIABLE CONTROL

STRATEGIES FOR DISTILLATION COLUMNS

C. O. Schwanke, T. F. Edgar, J. O. Hougen
Department of Chemical Engineering
The University of Texas at Austin
Austin, Texas

ABSTRACT

Modern multivariable techniques offer considerable promise in improving the quality of process control in chemical plants. While more complicated modeling exercises are required with these methods, they do provide controller synthesis strategies which directly treat the interactions among multiple inputs and outputs. In this work, several frequency domain methods have been tested using multivariable models from a pilot scale binary distillation column. The advantages of using multivariable control have been demonstrated in this study.

INTRODUCTION

Multivariable control techniques offer considerable promise in improving the quality of process control. In distillation processes, effective automatic process control can contribute to minimum energy usage by minimizing off specification products which must be reprocessed before sale, by operating closer to the desired product set points, and by increasing the recovery of desired product. The models used to design these controllers are generally more exact and account for multivariable interactions; hence, operations can be carried out with less variation in product quality and a lower controller effort. In addition, these multivariable techniques allow simultaneous feedback control of two or more outputs with as many controls while compensating for dynamic interaction. This means, for example, that the composition of both product streams from a binary distillation column may be controlled simultaneously, thus reducing the quantity of product which might be discarded.

Multivariable control theory has not yet been widely implemented in industrial operations. Several successful applications have appeared in the literature, but the number of failures is not known since they are seldom disclosed. Probably the major reason for its nonacceptance is the apparent complexity in structure and application; the fact that each multivariable control application requires more engineering design and trouble shooting time is another disadvantage.

In the implementation of multivariable control, the selection of the model or modeling procedure is most crucial. All mathematical models are in a sense approximations to the true behavior of the system studied. In the selection of a model for simulation and control, one must decide a priori what accuracy is needed in the model. The selection of the model and the implementation of the control are intimately related; for computer control the complexity of the model directly affects both the computer storage requirements and the computing time. This has often led to the formulation of so-called "black-box" models, which usually consist of simple dynamic relationships between inputs (controlled and uncontrolled) and outputs.

Generally, applications of multivariable control rely on the development of a physical model. Such a model for a binary distillation column was developed by Huckaba et al.[10], and is based on the following assumptions: negligible vapor holdup, negligible heat losses, constant holdup per plate, and uniform liquid composition on each tray. Material and energy balances for each plate plus reboiler and condenser form the ordinary differential equation describing the dynamic behavior of the system. For a binary mixture, $2(N + 2)$ equations result (N = number of trays). By expressing the enthalpy as a linear function of concentration, one can arrive at $(N + 2)$ nonlinear ordinary differential equations. This model can be further simplified by linearizing the differential equations about the desired steady state operating point, which then allows application of a number of established multivariable control design techniques. Data required either in equation form or tabular form include vapor-liquid equilibrium relationships, average heating capacity of subcooled liquid, boiling point data, and plate efficiencies as a function of composition. Plate hold-ups need to be estimated through dynamic testing, while plate efficiencies must be determined through steady state testing, as discussed by Oakley and Edgar[14].

As is apparent from the preceding paragraph, the development of a physical model for even a small column with two components is not a trivial exercise, and can be quite time-consuming. When attempting to perform a similar study on a multi-component industrial-scale column, it would be nearly impossible to obtain the requisite engineering data for model development. In addition, the number of differential equations increases proportional to the number of trays and the number of components. Hence it becomes clear that physical models often are not practical for control system design of large scale systems. However, physical model simulation packages can play a role in control system design, but usually for testing controllers developed from simpler models, such as black-box models obtained by input/output testing of the actual plant.

In this research the feasibility of applying modern control concepts using models obtained directly from experiments with the real plant has been examined. The models used in this study were derived by fitting the frequency response relations between single input-single output pairs of variables obtained from pulse tests of a pilot plant scale distillation column. This column is used to separate a binary mixture of methanol and water. These models were then converted to the various multivariable forms required for controller design. The frequency domain design methods studied included:

(1) classical single loop analysis
(2) decoupling control
(3) characteristic loci.

The application of these methods will be discussed in more detail in later sections of this paper. Time domain (state space) controller synthesis methods were also studied, including both continuous and discrete optimal feedback control; details of this research have been discussed by Schwanke et al.[19].

PREVIOUS CONTROL STUDIES

Most previous studies on the application of multivariable control to distillation columns have been devoted to simulation studies on pilot-plant experimental research e.g., Hu and Ramirez[9], and Rafal and Stevens[15]. A majority of this research has been based on state variable methods, which will not be discussed here; a thorough review of these applications has been performed by Schwanke[19]. It should be mentioned that many of these so-called multi-variable control studies have been concerned with control of a single output only (e.g., Heidemann and Esterson[6], Brosilow and Handley[5]. Experimental multiple input-multiple output control has been undertaken by several investigators. Scholander[17] has studied optimal feedback control of a pilot plant stripping column, while Oakley and Edgar[14] have researched optimal feedback control of a binary (methanol/water) distillation column. Binder and Calvillo[2] have also investigated optimal computer control of distillation column.

Frequency domain methods for multivariable systems (e.g. MacFarlane[13]) are of great interest presently, but most of these developments are so new that the reported distillation column applications have been limited to the well-known decoupling approach. Decoupling for multivariable column control has been studied by Toijala and Fagervik[21], Rijnsdorp and van Kampen[16], Luyben and Vinante[12], Stainthorpe and Jackson[20], and Wood and Berry[22].

EXPERIMENTAL SYSTEM DESCRIPTION AND TESTING PROCEDURE

The ten tray distillation column on which this work was based is shown in Figure 1. For this particular system there are five independent variables: the feed rate, F, the feed composition, x_F, the feed temperature, T_F, the steam pressure in the reboiler, P_{Stm}, and the reflux flow rate, L_{10}. The two product flow rates are used for controlling the inventory in the sump and accumulator. Hence they are fixed by the material balance.

Typically, in a large scale application one of the product compositions must be precisely controlled while the other product composition is not. However, operating costs could be reduced by regulating both compositions simultaneously. By controlling the bottoms composition as well as the overhead in methanol/water separation, one is able to regulate the amount of methanol which is discarded while maintaining the distillate composition at its desired composition. L_{10} and P_{Stm} were chosen as the control variables with which x_1 and x_D were controlled.

In the experimental system, x_1 and x_D were not directly measured. These compositions can be calculated from measurements of the bottoms liquid temperature (T_{1L}) and the overhead vapor temperature (T_{9V}). x_1 is calculated from bubble point data, x_D can be calculated at the corresponding dew point. Consequently, T_{1L} and T_{9V} are considered to be the output variables which are to be controlled.

The control problem is multivariable since two outputs are to be controlled by two inputs. In some cases two classically designed single loop controllers could be utilized, where, for example, L_{10} is used to control x_D while P_{Stm} controls x_1. However, a change in L_{10} affects x_1 as well as x_D and a change in P_{Stm} also affects both outputs. As will be seen, neglecting this interactive effect can yield undesirable results. Because of the interactive behavior of the system, multivariable controllers which account for interaction in their design should be considered for precise control.

All the models obtained from direct experimental tests in this work were obtained using the technique known as "pulse testing" (Hougen and Walsh[8]). This type of testing yields dynamic data which can be represented by frequency response diagrams, better known as Bode plots. The frequency response was then fitted to a Laplace domain transfer function of the desired order.

The following four SISO transfer functions resulted from the testing program.

$$G_{11} = \frac{T_{1L}}{P_{Stm}}(s) = \frac{1.84}{1.87s^2 + 2.19s + 1} \quad (1)$$

$$G_{12} = \frac{T_{1L}}{L_{10}}(s) = \frac{-28.14e^{-.65s}}{1.90s^2 + 2.21s + 1} \quad (2)$$

(time units in minutes)

$$G_{21} = \frac{T_{9V}}{P_{Stm}}(s) = \frac{.52(1 + 3.125s)}{1.78s^2 + 1.87s + 1} \quad (3)$$

$$G_{22} = \frac{T_{9V}}{L_{10}}(s) = \frac{-10.8(1 + 3.08s)}{2.130s^2 + 2.04s + 1} \quad (4)$$

Note that the only time delay occurs between reflux flow and the bottom temperature (T_{1L}). Figures 2 and 3 compare the time domain response of these systems with the actual smoothed experimental pulse tests. The agreement of the time histories of these simple models with the actual experimental data was quite good.

Because each input affects both outputs, the experimental model of the distillation column can be written in the following form:

$$\begin{bmatrix} y_1(s) \\ y_2(s) \end{bmatrix} = \begin{bmatrix} G_{11}(s) & G_{12}(s) \\ G_{21}(s) & G_{22}(s) \end{bmatrix} \begin{bmatrix} u_1(s) \\ u_2(s) \end{bmatrix} \quad (5)$$

where

$$y_1(s) = T_{1L}(s)$$
$$y_2(s) = T_{9V}(s)$$
$$u_1(s) = P_{Stm}(s)$$
$$u_2(s) = L_{10}(s)$$

Equation (5) is validated by expanding it into an equation for $y_1(s)$ and an equation for $y_2(s)$. These are given by

$$y_1(s) = G_{11}(s) u_1(s) + G_{12}(s) u_2(s) \quad (6)$$

$$y_2(s) = G_{21}(s) u_1(s) + G_{22}(s) u_2(s) \quad (7)$$

When pulse testing the column, u_2 was held constant at its steady state, while u_1 was varied, or vice-versa. Bode plots for $G_{11}(s)$, $G_{12}(s)$, $G_{21}(s)$, and $G_{22}(s)$ were obtained. Simultaneous pulsing of u_1 and u_2 to obtain the transfer functions makes it quite difficult to discriminate among possible models and to identify the unknown parameters.

SINGLE LOOP CONTROLLER DESIGN

A synthesis method developed by Hougen[7] was used to obtain PID controllers for a conventional single loop control configuration, i.e., the reflux controlling the overhead temperature and the steam pressure controlling the bottom temperature. An interactive computer program and CRT display was used to facilitate rapid synthesis of the controller constants. The essence of the design procedure used is as follows:

(1) Given the process frequency response curves, time constants are determined so that the system can be represented by

$$G_p(s) = \frac{(1+\tau_4 s)e^{-\tau_d s}}{(1+\tau_1 s)(1+\tau_2 s)(1+\tau_3 s)} \quad (8)$$

with appropriate selection of the τ's as needed.

(2) The electronic controller transfer function is given by

$$G_c(s) = K_c\left(\frac{1+T_i s}{T_i s}\right)\left(\frac{1+\alpha T_D s}{1+T_D s}\right) \quad (9)$$

where K_c, T_i, and T_D are the controller parameters and $\alpha=10$. A value is chosen for T_i so that when the controller is combined with the process, the slope of the amplitude data is -20 db per decade to as high a frequency as possible.

(3) T_D is chosen so that the frequency at which the process and controller phase angle sum to $-150°$ coincides with the frequency at which the maximum phase angle contributed by the derivative function occurs.

(4) The open loop frequency response is superimposed on a Nichols chart and the controller gain, K_c, necessary to make the open loop amplitude curve tangent to the 2 db locus is found.

(5) The closed loop frequency response is displayed and the crossover frequency is recorded.

(6) T_D is changed by a small amount, say 10%, and steps 4 and 5 are repeated. This procedure is repeated until the values of T_D and K_c which maximize ω_{co} are determined.

(7) T_i is decreased by a small amount and steps 3 through 6 repeated, thus obtaining new values of T_D, K_c, and ω_{co}. If ω_{co} is not greatly decreased, this step is repeated. A minimum value for T_i is desired, since this assures the most rapid integration for disturbance inputs without greatly retarding the closed loop time response. Execution of these steps yields the best controller parameters, subject to the performance criteria set forth previously.

A SISO controller ($G_{c_{11}}$) can thus be designed to control the bottoms composition via the steam pressure. The system model used for this design was $G_{11}(s)$. To control T_{9V} via the reflux flow rate simultaneously, the controller was conventionally designed with $G_{22}(s)$ as the system model ($G_{c_{22}}$), ignoring the interactions among the controllers and outputs. However, with the (u_1,y_1) loop closed, as shown in Figure 4, the transfer function between $L_{10}(s)$ and $T_{9V}(s)$ becomes much more complicated. Block diagram algebra was used to develop the following relationship between $L_{10}(s)$ and $T_{9V}(s)$:

$$\frac{y_2(s)}{u_2(s)} = \frac{T_{9V}(s)}{L_{10}(s)} = G_{22}(s) - \frac{G_{c_{11}} G_{12} G_{21}(s)}{1+G_{c_{11}} G_{11}(s)}$$

$$= G_{22}^*(s) \quad (10)$$

As discussed by MacFarlane[13], the term $-\frac{G_{c_{11}} G_{12} G_{21}(s)}{1+G_{c_{11}} G_{11}(s)}$ contributes to added phase lag. Neglecting this effect and designing on the basis of $G_{22}^*(s)$ is not altogether valid, since $G_{11}(s)$ becomes inaccurate when this loop is closed.

The error introduced by ignoring the interaction term in Equation (10) is shown by Figure 5, where the frequency responses of $G_{22}(s)$ and Equation (10) are compared. Note that the interaction terms have added phase lag characteristics at high frequencies to the original transfer function (G_{22}) and the gain has increased. Both of these effects significantly affect single loop design, thus emphasizing the need to predict the importance of interaction and to account for it in controller design. If the T_{9V} loop is closed and the transfer function relating $T_{1L}(s)$ to $P_{Stm}(s)$ is considered, the following results:

$$\frac{y_1(s)}{u_1(s)} = \frac{T_{1L}(s)}{P_{Stm}} = G_{11}(s) -$$
$$\frac{G_{12} G_{21} G_{c_{22}}(s)}{1+G_{c_{22}} G_{22}(s)} = G_{11}^* \quad (11)$$

The error introduced by using $G_{11}(s)$ for controller design instead of Equation (11) is shown in Figure 6. This Figure indicates that the same difficulties will occur for synthesis of $G_{c_{11}}$

When attempting to control both temperatures simultaneously using two single loop controllers, using the design procedure of Hougen[7], the controller matrix for $\underline{u} = \underline{\underline{G}}_c \underline{y}$ becomes

$$\underline{\underline{G}}_c(s) = \begin{bmatrix} G_{c_{11}}(s) & 0 \\ 0 & G_{c_{22}}(s) \end{bmatrix} \quad (12)$$

Figure 7 displays the Nyquist plot for the above controller matrix. The Nyquist plot for multivariable systems is based on a set of curves rather than a single curve.

For a given $G_{c_{22}}$, the controller for G^*_{11} (see equation (10)) can be checked for stability limits. By considering several $G_{c_{22}}$, the set of curves is constructed. Likewise, G^*_{22} (equation (10)) can be used to define a set of curves for testing stability limits of $G_{c_{22}}$ for various values of $G_{c_{11}}$. The curve for G^*_{11} very nearly encloses the -1.0 point, hence the stability margin was much smaller than predicted by single loop considerations. Figure 8 shows the Nyquist plot of the same system with both controller gains reduced by a factor of 0.1. The stability margin has been greatly increased, which illustrates the fact that single loop controllers often must be "detuned" before application to a multiple input-multiple output system.

Figures 9 and 10 illustrate the time domain response for the "detuned" single loop controllers discussed above. The effects of the system interactions are manifested in the excessive oscillations of both controllers, especially during the initial period of time. In Figure 9, for a setpoint change in only T_{11}, a major oscillation in T_{9V} occurs for τ.

The importance of interaction could have been predicted through the use of Bristol's interaction index [4]. He has defined a relative gain matrix, $\underline{\underline{M}}$, such that its elements m_{ij} are calculated from the following equation:

$$m_{ij} = \frac{\left(\frac{\partial y_i}{\partial u_j}\right)_{u_k}}{\left(\frac{\partial y_i}{\partial u_j}\right)_{y_k}} \quad (13)$$

where

y_i = the ith output variable

u_j = the jth control variable

y_k = all output variables held constant except the jth one.

u_k = all control variables held constant except the jth one.

Among the properties of $\underline{\underline{M}}$ listed by Bristol, the following defines the form $\underline{\underline{M}}$ takes on when interactive effects are significant and the integral mode is used in the controller:

"The transfer function between y_i and u_j...with all other loops closed will be nonminimum phase or unstable if m_{ij} is negative."

Hence, if negative elements exist in $\underline{\underline{M}}$, either a different control or output variable should be considered, or a controller which accounts for interaction in its design must be used. Juantorena and Romeo[11] have performed studies illustrating this fact.

For a 2 x 2 system, Juantorena and Romeo[11] have presented the following method for calculating $\underline{\underline{M}}$. First, $\underline{\underline{D}}$ is calculated from

$$\underline{\underline{D}} = \begin{bmatrix} \left(\frac{\partial y_1}{\partial u_1}\right)_{u_2} & \left(\frac{\partial y_1}{\partial u_2}\right)_{u_1} \\ \left(\frac{\partial y_2}{\partial u_1}\right)_{u_2} & \left(\frac{\partial y_2}{\partial u_2}\right)_{u_1} \end{bmatrix} \quad (14)$$

Then, define $\underline{\underline{C}}$ by the transpose of $\underline{\underline{D}}^{-1}$:

$$\underline{\underline{C}} = (\underline{\underline{D}}^{-1})^T \quad (15)$$

The elements of $\underline{\underline{M}}$ are now given by

$$m_{ij} = c_{ij} d_{ij} \quad (16)$$

Equation (16) can be proven equal to Equation (13) by algebraic and partial derivative manipulation.

For the column investigated in this work, the 2 x 2 version--Equations (14)-(16)--was used. $\underline{\underline{D}}$ is simply a matrix of steady state process gains which were given by

$$\underline{\underline{D}} = \begin{bmatrix} 1.842 & -28.14 \\ .52 & -10.8 \end{bmatrix} \quad (17)$$

Inverting and transposing $\underline{\underline{D}}$ yielded the following result for $\underline{\underline{C}}$:

$$\underline{\underline{C}} = \begin{bmatrix} 2.053 & .09884 \\ -5.349 & -.3501 \end{bmatrix} \quad (18)$$

$\underline{\underline{M}}$ was calculated by applying Equation (16), yielding

$$\underline{\underline{M}} = \begin{bmatrix} 3.782 & -2.782 \\ -2.782 & 3.782 \end{bmatrix} \quad (19)$$

Because $\underline{\underline{M}}$ contained negative elements, serious interactive effects were implied, as indicated by Bristol[4], thus emphasizing the need to include these effects in controller design.

In light of the difficulties with single loop control, several multivariable frequency domain design methods have been tested. A simulator based on a nonlinear physical model (Huckaba et al.[10]) has been used to test the multivariable controllers synthesized. This simulation has been successful in modeling the transient behavior (Oakley and Edgar[14]) for the column used in this study. Due to mechanical difficulties in the mini-computer system used, experimental validation of the controllers has not yet been completed.

DECOUPLING CONTROL

Decoupling is a controller design technique which decouples or eliminates system interaction in a conceptually straightforward manner. The technique is a well known design method, first presented by Boksenbom and Hood[3]. Given the system model in transfer function matrix form, $\underline{G}_p(s)$, a matrix controller, $\underline{G}_c(s)$, is chosen, the form of which diagonalizes the open loop transfer function matrix, $\underline{G}_{OL}(s)$. The open loop matrix is given by

$$\underline{G}_{OL}(s) = \underline{G}_p(s) \underline{G}_c(s) , \quad (20)$$

where $\underline{G}_c = \begin{bmatrix} G_{c_{11}} & G_{c_{12}} \\ G_{c_{21}} & G_{c_{22}} \end{bmatrix}$ and

$\underline{G}_p = \begin{bmatrix} G_{11} & G_{12} \\ G_{21} & G_{22} \end{bmatrix}$ and the closed loop

transfer function is given by

$$\underline{G}_{CL}(s) = [\underline{I} + \underline{G}_{OL}(s)]^{-1} \underline{G}_{OL}(s) \quad (21)$$

and $\underline{I}$ is the identity matrix.

When $\underline{G}_{OL}(s)$ is diagonal, $\underline{G}_{CL}(s)$ will also be diagonal, indicating the elimination of interaction. The controller parameters which appear in the diagonal elements are then adjusted to give the desired closed loop response.

For a system with two outputs, the diagonalization criterion yields relationships of the form,

$$G_{c_{21}}(s) = -G_{c_{11}}(s) \frac{G_{21}}{G_{22}}(s) \quad (22)$$

$$G_{c_{12}}(s) = -G_{c_{22}}(s) \frac{G_{12}}{G_{11}}(s) \quad (23)$$

These expressions can be substituted into the open or closed loop matrix form, Equations (20) and (21); transfer functions are chosen for $G_{c_{11}}(s)$ and $G_{c_{22}}(s)$ which satisfy some design criteria and then the forms of $G_{c_{12}}$ and $G_{c_{21}}(s)$ are

fixed by Equations (22) and (23). Equations (22) and (23) which define the off-diagonal terms can lead to complicated transfer functions. The complex form of the controller is one disadvantage of this technique.

One other significant disadvantage of decoupling is that positive zeroes in the transfer function can give rise to positive poles in the decoupler. Theoretical coupling is based on exact cancellation of poles and residues, hence the controller can be very unsatisfactory when model errors arise. However, exact decoupling is not necessary for effective multivariable control, and approximate decoupling may in fact yield superior results for multivariable control of an actual system.

For this two input-two output system the expressions which describe the off-diagonal elements of the controller have been given in Equations (22) and (23). Upon substitution of the actual transfer functions for the elements of the process transfer function matrix, these off-diagonal terms were given by:

$$G_{c_{21}}(s) = -G_{c_{11}}(s) \left[\frac{.52}{-10.8}\right] \left[\frac{1+3.12s}{1+3.08s}\right]$$
$$\left[\frac{1+2.04s+2.13s^2}{1+1.87s+1.78s^2}\right] \quad (24)$$

$$G_{c_{12}}(s) = -G_{c_{22}}(s) \left[\frac{-28.14}{1.84}\right]$$
$$\left[\frac{1+2.19s+1.87s^2}{1+2.1s+1.90s^2}\right] e^{-.65s} \quad (25)$$

The similarity between numerator and denominator transfer functions for this system were quite remarkable. It was assumed that the above rational fractions in s were sufficiently close to unity to allow cancellation, which resulted in the following expressions:

$$G_{c_{21}}(s) = -G_{c_{11}}(s)(-.04814815) \quad (26)$$

$$G_{c_{12}}(s) = -G_{c_{22}}(s)(-15.27687 e^{-.65s}) \quad (27)$$

The fact that the respective polynomials do not exactly cancel was the reason the final controller form does not truly represent non-interactive control. However, the resulting simplicity is certainly a desirable feature. The matrix controller was now defined to be:

$$\underline{G}_c(s) = \begin{bmatrix} G_{c_{11}} & (15.28 e^{-65s} G_{c_{22}}) \\ .04815 G_{c_{11}} & G_{c_{22}} \end{bmatrix} \quad (28)$$

where expressions for $G_{c_{11}}(s)$ and $G_{c_{22}}(s)$ remained to be chosen.

The different forms considered for $G_{c_{11}}$ and $G_{c_{22}}$ were: (1) pure proportional and (2) proportional plus integral. To obtain peak performance, derivative action might also be included, but was not in this work. Considering the proportional only case and using standard single loop methods for the decoupled closed loop control transfer function $\underline{G}_{CL}(s)$, values of 5.0 and 1.0 for $G_{c_{11}}$ and $G_{c_{22}}$ respectively yielded a system which was more oscillatory (especially in T_{9V}) than desired; however, lower gain values resulted in excessive steady state offset. The interactions appeared to be treated rather successfully. The controller matrix in this case was given by:

$$\underline{G}_c(s) = \begin{bmatrix} 5. & 15.27 e^{-65s} \\ .24075 & 1. \end{bmatrix} \quad (29)$$

When integral action was included in $G_{c_{11}}$ and $G_{c_{22}}$, the following transfer functions yielded satisfactory results:

$$G_{c_{11}} = 5 \frac{1+1.25s}{1.25s} \quad (30)$$

$$G_{c_{22}} = 0.6 \frac{1+.25s}{.25s} \quad (31)$$

The matrix controller was now given by

$$\underline{G}_c = \begin{bmatrix} 5\left(\frac{1+1.25s}{1.25s}\right) & 9.162 e^{-65s}\left(\frac{1+1.25s}{1.25s}\right) \\ .24075\left(\frac{1+1.25s}{1.25s}\right) & .6\cdot\left(\frac{1+1.25s}{1.25s}\right) \end{bmatrix}$$

$$(32)$$

The closed loop frequency response for this system is shown in Figure 11. As a result of the integral action, the amplitude ratios of the diagonal terms approached 0 db as $\omega \rightarrow 0$, thus eliminating steady state offset. T_{1L}/T_{1L_r} had a peak amplitude ratio of roughly 3 db while T_{9V}/T_{9V_r} peaked at 5 db. Both were larger than the desired values for design of a single loop controller. If desired, the magnitudes of these peaks could be reduced by decreasing the proportional gains in $G_{c_{11}}$ and $G_{c_{22}}$ at the expense of slower response to set point changes. The frequency response of the off-diagonal terms indicated that interaction was successfully limited via the approximate form of decoupling. Due to the integral action the amplitude ratios for these off-diagonal terms approached $-\infty$ db as $\omega \rightarrow 0$. Hence steady state interaction between the two control loops was totally eliminated.

System responses to set point changes are shown in Figures 12 and 13. The inclusion of integral action made it possible to decrease the proportional gain in $G_{c_{22}}$, which greatly reduced the oscillatory response of T_{9V} observed in the decoupler with proportional action only.

CHARACTERISTIC LOCI METHOD

This technique developed by A.G.J. MacFarlane[13] extends the single loop stability criteria of Bode and Nyquist to multivariable systems. Utilizing a system model in transfer function matrix form, a stability criteria is defined and a systematic procedure for design of a matrix controller is followed, as discussed by Belletrutti[1]. It is a trial and error technique and must be implemented on an interactive computer system as was the SISO design.

$\underline{G}_c(s)$ must simultaneously perform many functions. It must modify the characteristic loci in order to satisfy the stability criteria, achieve acceptable interaction, and meet pre-defined performance criteria. It is useful then to let $\underline{G}_c(s)$ be the product of several matrix sub-controllers so that these can be designed to handle the above tasks one at a time. $\underline{G}_c(s)$ is then given by

$$\underline{G}_c(s) = \prod_{i=1}^{q} \underline{G}_c^{i}(s) \quad (33)$$

The restrictions on the $\underline{G}_c^{i}(s)$ as given by Belletrutti[1] are as follows:

(1) each $\underline{G}_c^{i}$ should be as simple as possible,
(2) all the dynamic elements of $\underline{G}_c^{i}(s)$ should be rational functions,
(3) $\underline{G}_c^{i}(s)$ should be nonsingular,
(4) all poles of $\underline{G}_c^{i}(s)$ must lie in the open left half plane, and
(5) the determinant of $\underline{G}_c^{i}(s)$ must not have any right-half plane zeroes.

Belletrutti has also presented several useful forms for $\underline{\underline{G}}_c^1(s)$ and their effects on the system's characteristics.

In the design procedure, the number of right-half plane zeroes of the open loop characteristic polynomial was determined first. Then the closed loop stability using Nyquist plots was tested with $\underline{\underline{G}}_c(s)$ initially equal to the identity matrix $\underline{\underline{I}}$. Due to the positive feedback which resulted from the negative gain of G_{22}, the system was only stable for $k_1 < .0604$ and $k_2 < 17.7$, where the following equation defines k_1 and k_2:

$$\underline{\underline{G}}_{OL}(s) = \underline{\underline{G}}_p(s) \, \underline{\underline{G}}_c(s) \begin{bmatrix} k_1 & 0 \\ 0 & k_2 \end{bmatrix} \cdot \underline{\underline{I}} \quad (34)$$

and $\underline{\underline{G}}_c$ equals the current composite controller matrix. These stability margins were improved by eliminating this positive feedback via the following sub-controllers:

$$\underline{\underline{G}}_c^1 = \begin{bmatrix} 1 & 0 \\ 0 & -1 \end{bmatrix} \quad (35)$$

which replaces the identity matrix ($\underline{\underline{I}}$) in Equation (34). The system was now stable for $k_1 > 0$ and $k_2 < 17.7$.

After a considerable stability margin was established, the amount of interaction present was tested. Interactive effects were investigated via the frequency response of $\dfrac{T_{1L}}{T_{9V_r}}(s)$ and $\dfrac{T_{9V}}{T_{1L_r}}(s)$, the off-diagonal terms of the closed loop matrix. The interaction frequency response plot for the system in question is displayed in Figure 14. The frequency response of $\dfrac{T_{1L}}{T_{9V_r}}$ clearly showed the presence of excessive interaction at low frequencies, since its steady state gain was +3.5 db. As recommended by Belletrutti, the "D.C. plant inverse controller matrix" was implemented to eliminate this low frequency interaction. Hence, $\underline{\underline{G}}_c^2$ was given by

$$\underline{\underline{G}}_c^2 = \begin{bmatrix} 2.053 & -5.349 \\ -.09884 & .3501 \end{bmatrix} \quad (36)$$

This updated system was tested for stability via the root loci. Stability was assured although the stability margin had decreased, with $k_1 < 4$ and $k_2 > 0$. Figure 15 demonstrates the successful elimination of low frequency interaction; however, considerable interaction for $\dfrac{T_{1L}}{T_{9V_r}}$ remained in the midfrequency range.

No sub-controller which conformed to the rules given by Belletrutti[1] was found which successfully limited the remaining interaction. Consequently, the performance phase of the design was considered. In order to increase the closed loop steady state gain, integral action was incorporated in the final sub-controller; after some iteration, the final controller matrix was

$$\underline{\underline{G}}_c = \begin{bmatrix} 1.0264\left(\dfrac{1+s}{s}\right) & -5.349\left(\dfrac{1+s}{s}\right) \\ -.04942\left(\dfrac{1+s}{s}\right) & .35014\left(\dfrac{1+s}{s}\right) \end{bmatrix} \quad (37)$$

This system's response to the set point changes described earlier is shown in Figures 16 and 17. Its response was slow, and considerable interactive effects were present compared to the response characteristics of the decoupling approach. Figure 16 shows the undesirable oscillations in T_{1L}, which appear to be more severe than for the single loop designed controllers (Figures 9 and 10).

CONCLUSIONS

Single loop and multivariable control methods which operate in the frequency domain have been studied for application to distillation column control. It was found that for this pilot-scale column serious interaction among outputs and controls limited the effectiveness of single loop methods. Single loop control, as analyzed through root locus diagrams and the interaction matrix technique, was unsatisfactory. Thus two multivariable controller design techniques were applied. The characteristic locus method was not successful in limiting the system interaction, possibly due to inexperience in application of this method. An approximate decoupling multivariable control was also applied, and it resulted in a controller of very simple form. The closed loop responses for this controller were quite satisfactory, showing quick response to set point changes and exhibiting very little interaction in the responses. It is felt that further examination of results from the decoupling approach may give some guidance for improvement of the characteristic loci method.

REFERENCES

(1) Belletrutti, J. J., "Computer-Aided Design and the Characteristic Loci Method," *IEE Conf. Publ.* 96, 79 (1973).

(2) Binder, Z., and Calvillo, L., "Dynamic Control of a Pilot Distillation Column by Digital Computer," Proc. 4th IFAC/IFIP Conf. on Digital Computer Applications to Process Control, Zurich (1974).

(3) Boksenbom, A. S., and Hood, R., "General Algebraic Method Applied to Control Analysis of Complex Engine Types," NACA-TR-930, Washington, D.C., 1949.

(4) Bristol, E. H., "On a Measure of Interaction for Multivariable Process Control," *IEEE Trans. Auto. Cont.*, AC-11, 133 (1966).

(5) Brosilow, C. B., and Handley, K. R., "Optimal Control of a Distillation Column," *AIChE J.*, 14, 467 (1968).

(6) Heidemann, R. A., and Esterson, G. L., "An Optimal Discrete Controller for a System with Load Changes," *Proc. JACC*, 454 (1967).

(7) Hougen, J. O., *Measurements and Control Applications for Practicing Engineers*, Barnes and Noble, 1972.

(8) Hougen, J. O., and Walsh, R. A., "Pulse Testing Method," *CEP*, 57, 69 (1961).

(9) Hu, Y. C., and Ramirez, W. F., "Application of Modern Control Theory to Distillation Columns," *AIChE J.*, 18, 279 (1972).

(10) Huckaba, C. E., May, E. P., and Franke, F. R., "An Analysis of Transient Conditions in Continuous Distillation Operations," *CEP Symp. Ser.*, 59, No. 46, 38 (1963).

(11) Juantorena, R., and Romeo, R. T., "Application of the Relative Gain Matrix to a Distillation Column," *Instrumentation in the Chemical and Petroleum Industries*, 7, 53 (1971).

(12) Luyben, W. L., and Vinante, C. D., "Experimental Studies of Distillation Decoupling," *Kem. Teollisuus*, 29, 499 (1972).

(13) MacFarlane, A. G. J., "A Survey of Some Recent Results in Linear Multivariable Feedback Theory," *Automatica*, 8, 455 (1972).

(14) Oakley, D. O., and Edgar, T. F., "Optimal Feedback Control of a Binary Distillation Column," *Proc. JACC*, 523 (1976).

(15) Rafal, M. D., and Stevens, W. F., "Discrete Dynamic Optimization Applied to On-line Optimal Control," *AIChE J.*, 14, 85 (1968).

(16) Rijnsdorp, J. A., and van Kampen, J. A., "Automatic Feedback Control of Two Product Qualities of a Distillation Column," *Proc. Third IFAC Congress*, London, 1966.

(17) Scholander, P., "Computer Control of a Pilot Plant Stripping Column," Proc. 4th IFAC/IFIP Conf. on Digital Computer Applications to Process Control," Zurich (1974).

(18) Schwanke, C. O., "Experimental Multivariable Control of a Pilot Plant Scale Distillation Column," Ph.D. Dissertation, The University of Texas at Austin, Austin, Texas, 1977.

(19) Schwanke, C. O., Edgar, T. F., and Hougen, J. O., "Problems in the Application of Optimal Multivariable Control Theory to Distillation Columns," Annual AIChE Meeting, Chicago, November, 1976.

(20) Stainthorpe, F. P., and Jackson, C. B., "Control of a Fractionating Column to Product Rate Demand Changes," *Proc. Third IFAC Symp. on Multivariable Technol. Systems*, Manchester, U.K., 1974.

(21) Toijala, K., and Fagervik, K., "A digital simulation study of two-point feedback control of distillation columns," *Kem. Teollisuus*, 29, 1 (1972).

(22) Wood, R. K., and Berry, M. W., "Terminal Composition Control of a Binary Distillation Column," *Chem. Engr. Sci.*, 28, 1707 (1973).

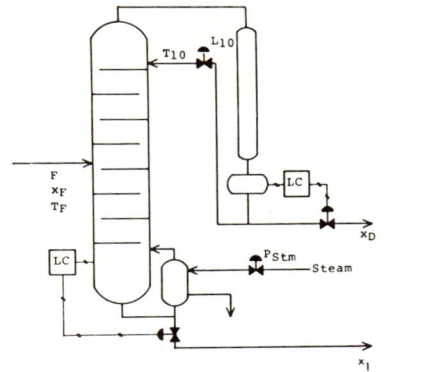
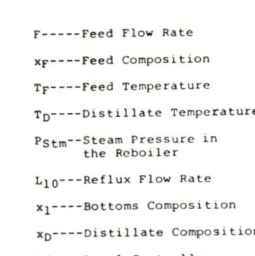

F-----Feed Flow Rate
x_F----Feed Composition
T_F----Feed Temperature
T_D----Distillate Temperature
P_{Stm}--Steam Pressure in the Reboiler
L_{10}---Reflux Flow Rate
x_1----Bottoms Composition
x_D----Distillate Composition
LC----Level Controller

Figure 1: Simplified Flow Chart of the Pilot Scale Distillation Column

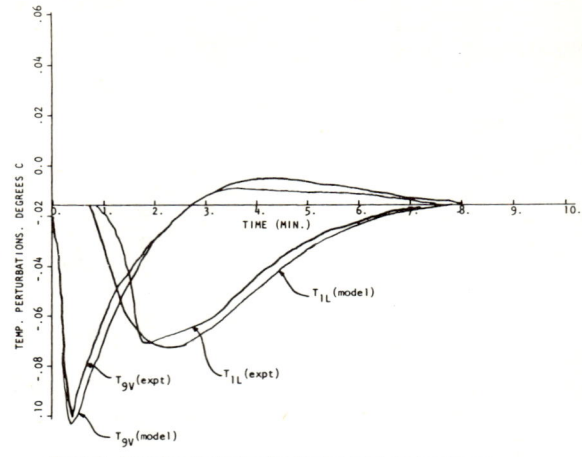

Figure 2: Comparison of the Transfer Function Model and Experimental Responses to a Reflux Pulse

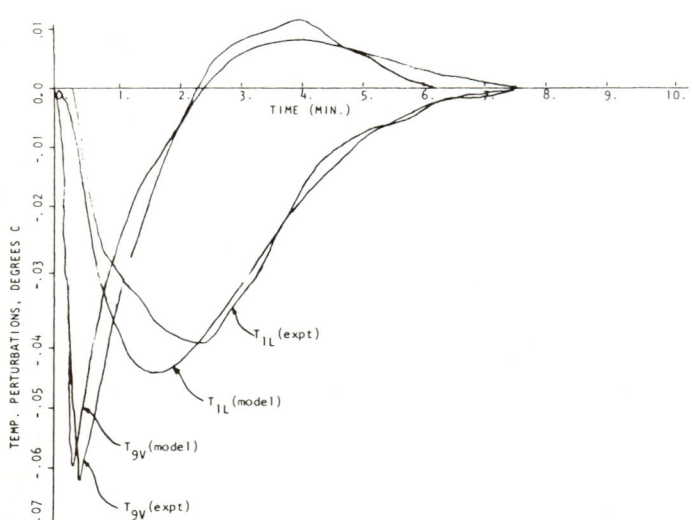

Figure 3: Comparison of the Transfer Function Model and Experimental Responses to a Steam Pulse

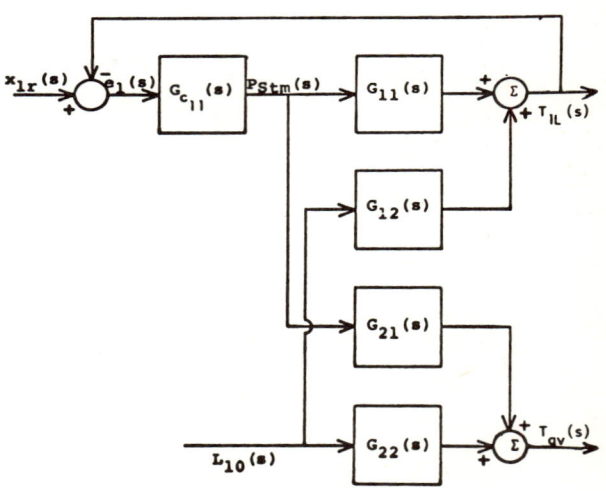

Figure 4: Block Diagram Model of the Distillation Column with the Bottoms Composition being Controlled by the Steam Pressure

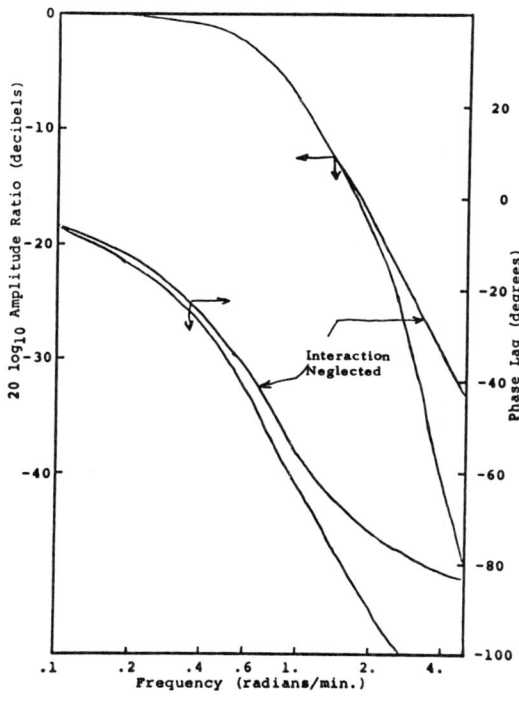

Figure 5: Effect of Interaction on $\frac{T_{9v}}{L_{10}}$ (s)

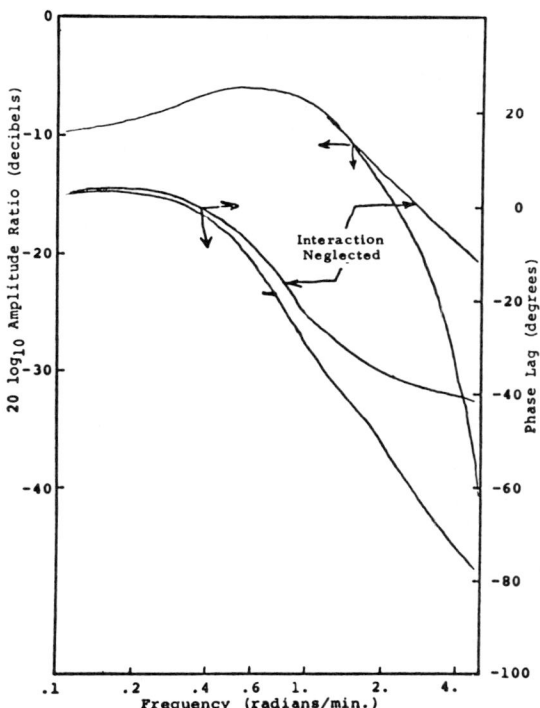

Figure 6. Effect of Interaction on $\frac{T_{1L}}{P_{stm}}$ (s).

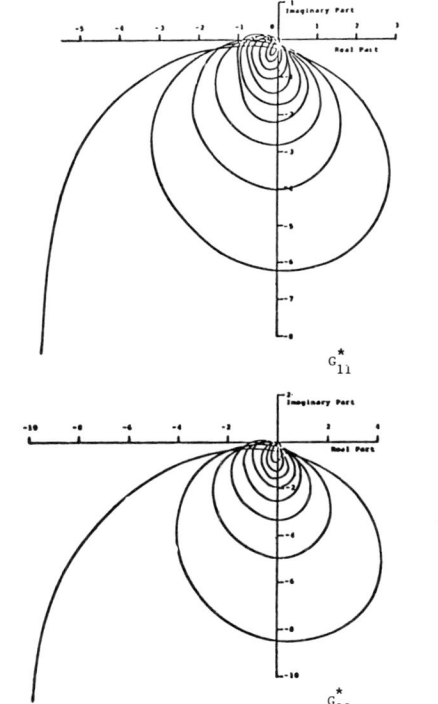

Figure 7: Root Loci of the Multivariable System when the Single Loop Controllers were Simultaneously Implemented

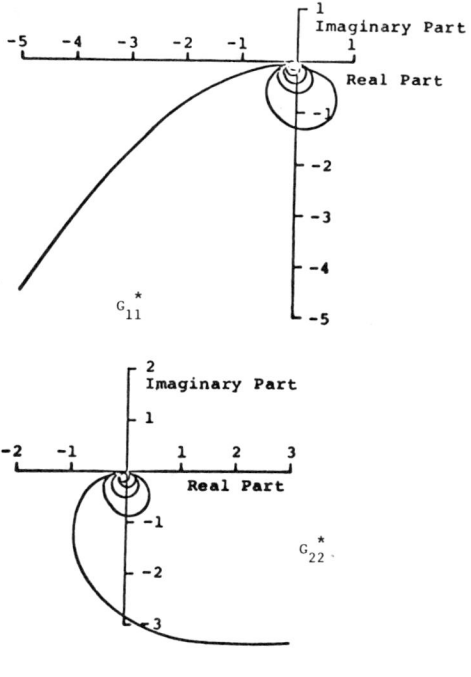

Figure 8: Root Loci of the Multivariable System when the Single Loop Controllers are Implemented with their Proportional Gains Reduced by a Factor of .1

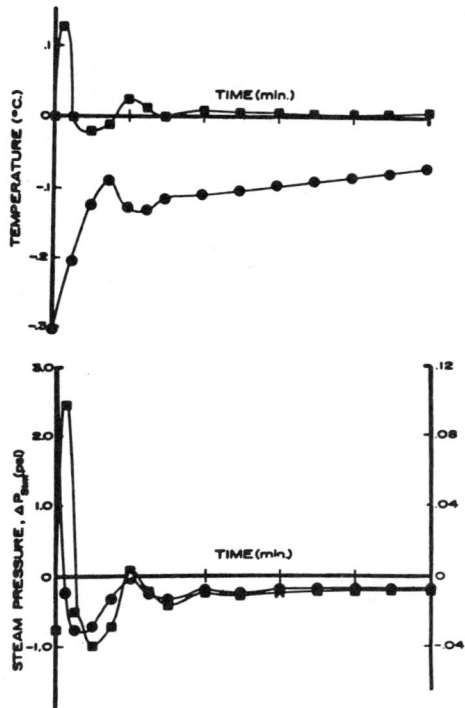

Figure 9: Multivariable Response to an Upset in the Initial Condition of T_{1L} using Simultaneous Single Loop Controllers

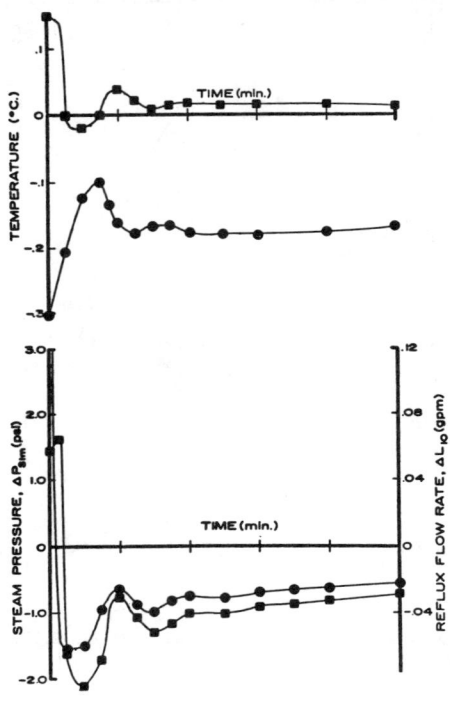

Figure 10: Multivariable Response to an Upset in the Initial Conditions of T_{1L} and T_{9v} Using Simultaneous Single Loop Controllers

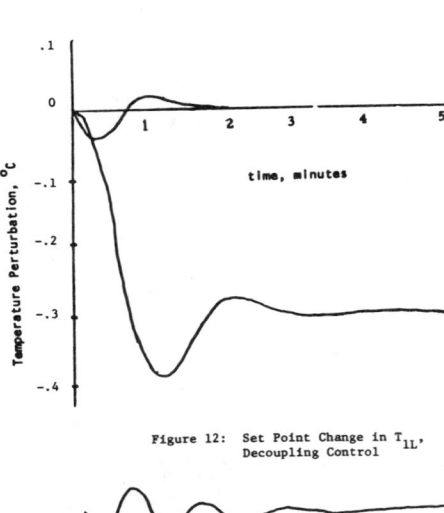

Figure 12: Set Point Change in T_{1L}, Decoupling Control

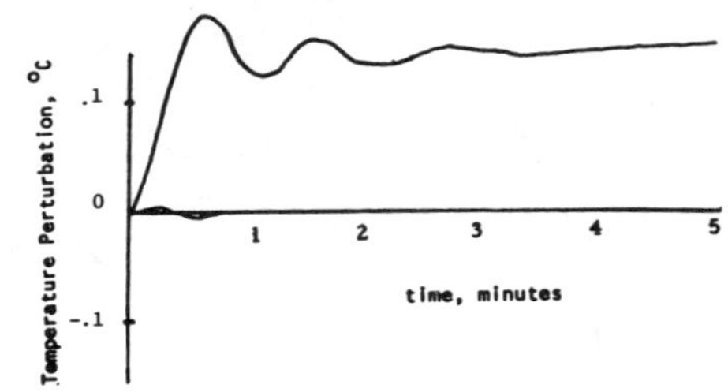

Figure 13: Set Point Change in T_{9v} Decoupling Control

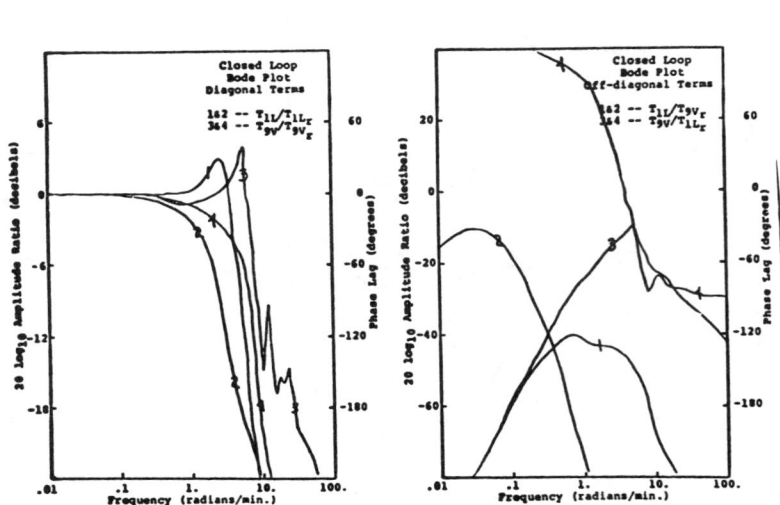

Figure 11: Closed Loop Frequency Response of the Proportional Plus Integral, Decoupling Controller

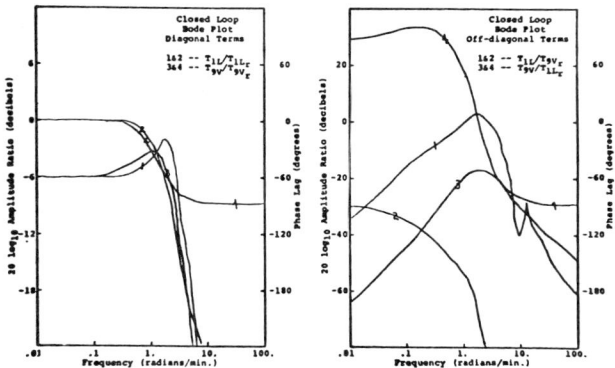

Figure 15: Closed Loop Frequency Response After Implementation of Equation (36)

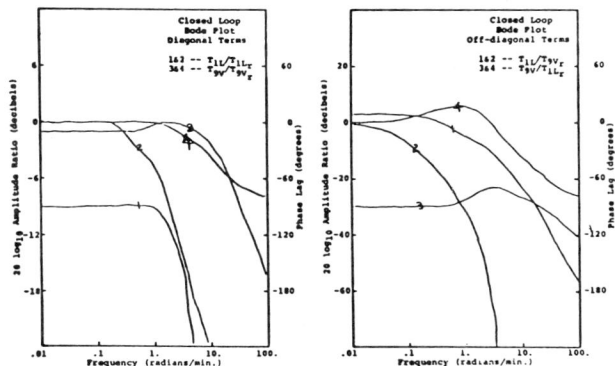

Figure 14: Closed Loop Frequency Response when the Controller given by Equation 34 is Implemented

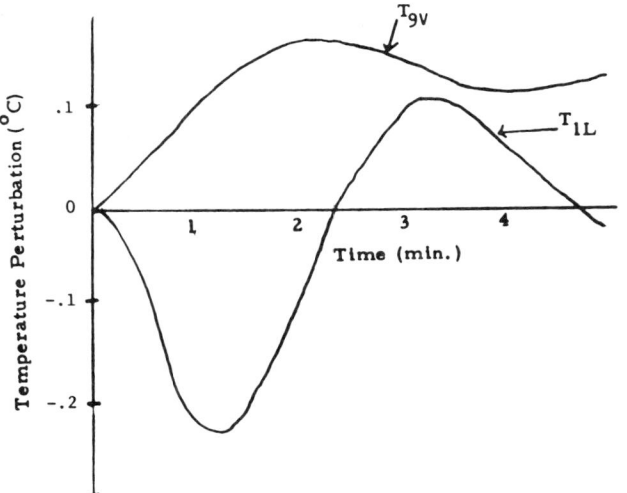

Figure 16: Closed Loop Response to a Set Point Change in T_{9V}; Characteristic Loci Controller

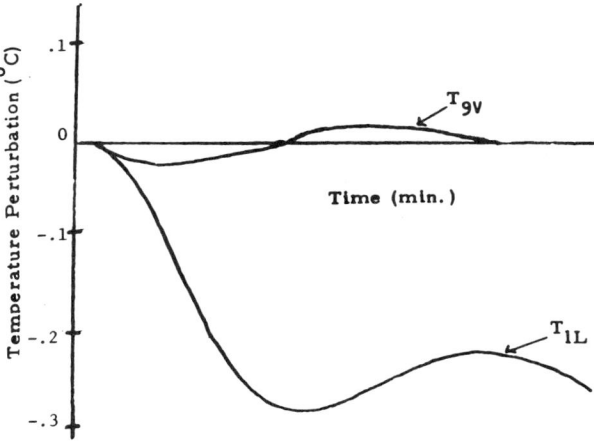

Figure 17: Closed Loop Response to a Set Point Change in T_{1L}; Characteristic Loci Controller

© ISA, 1977
ISBN 87664-363-2

APPLICATION OF A SIMPLE DISCRETE MULTIVARIABLE STRATEGY
TO A LABORATORY SCALE AUTOCLAVE

Buck Sayers
Computer Engineer
E. I. duPont de Nemours
Corpus Christi, Texas

Charles F. Moore
Professor
Department of Chemical and Metallurgical Engineering
The University of Tennessee
Knoxville, Tennessee

ABSTRACT

Linear regression analysis is applied in experimental multivariable process modeling of a laboratory scale batch autoclave. Using the apriori form of a first order matrix difference equation, design of a simple discrete first order decoupler is presented. The decoupler is compensated for time varying and nonlinear properties inherent in batch chemical processes. Experimental evaluation of the final non-interfacing, PID based control strategy is presented.

INTRODUCTION

The overall objective of this work is to design a practical nonlinear, multivariable, computer based control strategy for use on typical industrial reactors. The purpose of this strategy is to provide effective and non-interacting control of the various process variables and to maintain independent control in spite of the exaggerated nonlinear nature of some chemical processes such as batch reactors. The problem is divided into three general parts: (1) multivariable process modeling, (2) multivariable control strategy and (3) experimental evaluation of the closed loop control system.

Earlier work (18, 19) has shown the effectiveness of the discrete multivariable model and decoupler. This paper extends these techniques to experimental evaluation on laboratory equipment representative of actual chemical processing units.

BACKGROUND

Most chemical processes are inherently multivariable systems. A characteristic of these processes is that they exhibit interactions between the inputs and outputs as shown in Figure 1. Consequently when standard process control strategy is applied, the control loops are not independent as shown in Figure 2. The objective of multivariable control strategy is to compensate for the interactions and make the control loops independent. Several strategies are present in the references. The strategy which is extended here makes use of a discrete decoupler to reduce the process to independent loops as shown in Figure 3.

This approach to multivariable controller design is beneficial because it is easy to understand, allows the process industry PID controllers on multivariable systems and does not require block diagrams or a specific structural model. The model used in this work is termed nonspecific because it is generally applicable to most chemical processes. It is as general as the old familiar first order lag coupled with dead time.

For a multivariable process this model is

$$[\hat{R}]_{i+1} = [A][R] + [B][W]_i$$

where

$[\hat{R}]_{i+1}$ = vector of the predicted values of the N process responses at the next sample.

$[R]_{i+1}$ = vector of the N actual process responses measured at the current sample (time = i).

$[W]_i$ = vector of N process input at sample i.

$[A],[B]$ = constant matrices which are unique to this process but not analytically related to the transfer functions. The elements a_{ij} and b_{ij} are to be experimentally evaluated by linear regression analysis.

A simple first order discrete decoupler can be designed from this process model.

DESIGN OF THE COUPLER

From past experience with analog decoupling it is known that the process input [W] is a function of the controller output [M] and the decoupler output [Z]

$$[W]_i = [Z]_i + [M]_i$$

Substituting into the process model:

$$[\hat{R}]_{i+1} = [A][R]_i + [B][Z]_i + [B][M]_i$$

For a specific response $\hat{r}^R_{i+1}$ we have

$$\hat{r}^R_{i+1} = \sum_{j=1} a_{RJ}\, r^j_i + \sum_{j=1}^{N} b_{RJ}\, m^j_i + \sum_{j=1}^{N} b_{Rj}\, Z^j_i \quad (1)$$

For non-interaction Equation (1) requires

$$\sum_{j=1}^{N} b_{Rj}\, Z^j_i = -\sum_{\substack{j=1 \\ j \neq R}}^{N} a_{Rj}\, r^j_i - \sum_{\substack{j=1 \\ j \neq R}}^{N} b_{Rj}\, m^j_i \quad (2)$$

Equation (2) specifies the design of the decoupler whenever the elements of the [A] and [B] matrices are known. However, chemical process dynamics are rarely known. They usually must be approximated by experimental modeling.

EXPERIMENTAL MODELING

For a chemical process expressed as an analog p-canonical model it is well known how to model each $G_o(s)$ transfer function. For the V or modified V model it is not so obvious how one can experimentally determine the transfer functions.

The discrete model eliminates the modeling problem associated with the V and modified V models. The modeling begins with the model expressed as:

$$[\hat{R}]_{i+1} = [A][R]_i + [B][W]_i$$

The [A] and [B] matrices are to be determined by regression analysis on experimental responses data. The data are collected by a digital computer sampling the process outputs, [R], while varying the process inputs [W]. The objective is to calculate the [A] and [B] matrices such that the sum of the error squared at each sample is minimized, i.e.,

$$([\hat{R}]_1 - [R]_1)^2 + ([\hat{R}]_2 - [R]_2)^2$$
$$+ \ldots ([\hat{R}]_i - [R]_i)^2$$

is a minimum. Letting

$$F = \sum_{i=1}^{I} ([\hat{R}]_i - [R]_i)^2$$

we can state the necessary conditions for a minimum F as:

$$\frac{\partial F}{\partial a_{ij}} = \frac{\partial F}{\partial b_{ij}} = 0$$

This must hold for i = 1, 2, 3 . . . N and j = 1, 2, 3 , , , N. This is now in the form required for standard linear regression analysis.

DESCRIPTION OF EQUIPMENT

In the University of Tennessee's Department of chemical and Metallurgical Engineering a program exists to investigate digital computer control of chemical processes. A complete data acquisition and control system using a PDP 15/35 as its nucleus is in operation. Using the PDP-15 specific control strategies are implemented on laboratory scale equipment of VIA hybrid simulation of process equipment.

The laboratory units are built for research into practical control problems in the chemical industry. This paper investigates multivariable digital control of a laboratory scale batch autoclave.

The autoclave has an internal cylindrical body which is three feet high and fifteen inches in diameter. This internal body is fitted with an internal steam heating coil and is also steam jacketed. The autoclave is illustrated in Figure 4 along with the appropriate process variables. The controlled variables are usually two of the following:

Body Pressure
Liquid Temperature
Vent Flow Rate
Steam Pressure
Steam Flow
Liquid Composition

In this study the body pressure and the vent rate are selected as the primary control variables since these have the strongest dynamic interrelationships. The body pressure and vent rate are created by evaporating water using saturated steam as our heat source.

LABORATORY AUTOCLAVE STUDY

The discrete decoupler developed for linear multivariable systems cannot be applied directly to the laboratory autoclave. There are several nonlinearities which must be compensated. One of these nonlinearities is the control valves themselves. The nonlinearity associated with the steam control valve causes the linear decoupler to calculate incorrect steam valve positions. The result is incomplete decoupling.

One approach to eliminate this nonlinearity is to define a pseudo-process in which the steam flow is a process input. This is done by using a local controller on the steam flow. The concept of a pseudo-process is important because the local steam controller is actually a computer based algorithm, but is considered to be a fundamental part of the multivariable process. Figure 5 illustrates the distinction between the pseudo-process and the actual physical process. Since the dynamics of the steam loop are much faster than the process, the process effectively "sees" a step change in steam flow when we input a step in the steam flow set point. The controller eliminates any changes in steam flow due to pressure changes, valve hysteresis and decreases the process time constant.

The nonlinearities associated with the vent valve can also be eliminated with a local controller. The dynamics of the vent rate loop are so fast that the local controller can practically eliminate any vent rate fluctuations in spite of body pressure fluctuations. This local controller can also compensate for disturbances which enter via the body pressure and thus maintains the vent rate decoupled from the body pressure setpoint.

Since this is a batch process its transfer functions are expected to vary with time. These variations are caused by changes in the mass of the tank contents, and because the thermodynamic properties of the tank contents vary with time. The time variations due to the changing mass are compensated for by controlling the steam flow. Thus, at a constant steam setpoint the same amount of heat will be delivered, regardless of the tank level. The variations due to changing properties can be eliminated by experimentally establishing the relation between the [A] and [B] matrix elements and body pressure. Assuming that this relation is linear, only two experimental points are needed to establish the relation. The linearity can then be confirmed by trying to decouple at another point. The linear relation is shown in Table I. The success in decoupling confirms this linear relation throughout the complete range of liquid level. Figure 6 illustrates the block diagram of this strategy.

EVALUATION OF THE EXPERIMENTAL MODEL AND DECOUPLER

The pseudo-process model can be expressed as

$$\begin{bmatrix} BP \\ VR \end{bmatrix}_i = \begin{bmatrix} a_{11} & a_{12} \\ a_{12} & a_{22} \end{bmatrix} \times \begin{bmatrix} BP \\ VR \end{bmatrix}_{i=1} + \begin{bmatrix} b_{11} & b_{12} \\ b_{21} & b_{21} \end{bmatrix} \times \begin{bmatrix} SF \\ VV \end{bmatrix}_{i=1}$$

where each a_{ij} and b_{ij} is evaluated by linear regression analysis of process step response data.

As shown in Figure 6, the vent rate will be "decoupled" with a local controller. Thus, the discrete decoupler must only decouple the interactions caused by the vent rate setpoint and the vent valve opening. This reduces our model to

$$BP_{i+1} = a_{11}BP_i + a_{12}VR + b_{11}SF + b_{12}VV$$

Using the decoupler the process input is:

$$SF = CA + DA$$

where

CA = steam flow setpoint called for by the controller.

DA = steam flow setpoint called for by the decoupler.

The decoupled model becomes

$$BP_{i+1} = a_{11}BP_i + a_{12}VR_i + b_{12}VV_i + b_{11}CA_i + b_{11}DA$$

where

$$DA_i = -\frac{a_{12}VR_i + b_{12}VV_i}{b_{11}}$$

EXPERIMENTAL RESULTS

Using the system diagrammed in Figure 7, the data for the experimental modeling are obtained via step responses from the pseudo-process. As shown in Figures 8 and 9 these step responses illustrate a strong degree of interaction. The data from the step response are used in the regression analysis to determine the process matrices. These matrices are then used to calculate the elements of the decoupler. The effectiveness of the decoupler is tested with a step input to the open loop decoupled system as shown in Figure 10. These open loop decoupled responses are shown in Figures 11-13. The final objective is to test the closed loop multivariable control as shown in Figure 6. These responses are shown in Figures 14-17.

CONCLUSIONS

This work illustrates the application of a simple multivariable strategy which utilizes standard PID controllers in the feedback loops. The feedback loops are maintained independent by a simple first order matrix difference equation decoupler. The decoupler combines feedforward and feedback action and is "transparent" to the PID controller.

This effective multivariable strategy illustrates the application of a first order multivariable difference equation as a multivariable process model using linear regression analysis to evaluate the matrix elements. This model may be as general as the old reliable first order lag coupled with dead time.

Using simple cascade control loops to remove process nonlinearities a practical multivariable strategy can be based on the discrete decoupler. The simplicity of the technique may make it much more acceptable to the process industry as a multivariable strategy.

REFERENCES

1. Beard, C. D., *Final Control Elements*, Rimbach Publications, Division of Chilton Company, Philadelphia, 1969.

2. Chatterjee, H. K., "Multivariable Process Control," *Proc. First IFAC Conf.*, 1960, pp. 132-141.

3. Foster, R. D. and W. F. Stevens, "A Method for the Noninteracting Control of a Class of Multivariable Systems," *AIChE Journal*, Vol. 13, No. 2, March 1967, pp. 334-39.

4. Foster, R. D. and W. F. Stevens, "An Application for the Noninteracting Control of a Class of Multivariable Systems," AIChE Journal, Vol. 13, No. 2, March 1967, pp. 340-

5. Freeman, H., "A Synthesis Method for Multiple Control Systems," Trans. AIEE (Appl. and Ind.), Vol. 7, Pt. 2, March 1957, pp. 28-31.

6. Graupe, D., B. H. Swanick and G. R. Cassir, "Reduction and Identification of Multivariable Processes Using Regression Analysis," Trans. IEEE, October 1968, pp. 564-567.

7. Frank, Roger G. E., Mathematical Modeling in Chemical Engineering, Wiley, 1967.

8. Greenfield, G. G. and T. S. Ward, "Structural Analysis for Multivariable Process Control," I&EC Fundamentals, Vol. 6, No. 4, November 1967, pp. 571-580.

9. Greenfield, G. G. and T. J. Ward, "Feedforward and Dynamic Uncoupling Control of Linear Multivariable Systems," AIChE Journal, Vol. 14, No. 5, September 1968, pp. 783-789.

10. Lee, T. H., G. E. Adams and W. M. Gaines, Computer Process Control: Modeling and Optimization, John Wiley and Sons, Inc., New York, 1968.

11. Lloyd, S. G., "Basic Concepts of Multivariable Control," Instrumentation Technology, December 1973, pp. 31-36.

12. Mesarovic, M. D., The Control of Multivariable Systems, MIT Technology Press, 1960.

13. Minn, H. S. and T. J. Williams, "Chemical Process Control in the Presence of Both Transport Lag and Sampled-Data Control," Process Dynamics and Control, Chemical Engineering Progress Symposium Series, No. 36, 1961, p. 100.

14. Moore, C. F., Ph.D. Dissertation, Louisiana State University, 1969

15. Murrill, P. W., Automatic Control of Processes, International Textbook, Pennsylvania, 1967.

16. Pierini, P. E., M.S. Thesis, The University of Tennessee, 1971.

17. Sayers, H. D., Ph.D. Dissertation, The University of Tennessee, 1975.

18. Sayers, H. D. and C. F. Moore, "A Simple Multivariable Decoupler for DDC," ISA Annual Conference, October 1973.

19. Smith, C. L., Digital Computer Process Control, International Textbook Company, 1972.

20. Zalkind, C. S., "Practical Approach to Noninteracting Control, Part I," Instruments and Control Systems, Vol. 40, March 1967, pp. 89-94.

21. Zalkind, C. S., "Practical Approach to Noninteracting Control, Part II," Instruments and Control Systems, Vol. 40, April 1967, pp. 111-116.

NOMENCLATURE

K = process gain

τ = process time constant.

s = Laplace variable.

$w(s)$ = vector of Laplace transformed process input.

$R(s)$ = vector of Laplace transformed process output.

$[W]_i$ = vector of process inputs at sample i.

$[R]_{i+1}, [R]_i$ = vector of process inputs at samples i+1 and i, respectively.

$[A], [B]$ = constant matrices.

$[M]_i$ = vector of controller outputs at sample i.

$[Z]_i$ = vector of decoupler outputs at sample i.

$\hat{r}^h_{i+1}$ = k^h element of $[\hat{R}]_{i+1}$.

r^j_i = j^{th} element of $[R]_i$.

z^j_i = j^{th} element of $[Z]_i$.

m^j_i = j^{th} element of $[M]_i$.

N = number of inputs and outputs in the multivariable system.

a_{ij}, b_{ij} = elements of the ith row and j^{th} column of matrices [A] and [B], respectively.

F = error function, vector of errors $\hat{r}^k_{i+1} - r^k_{i+1}$.

T = total number of samples taken during the transient response.

f^k = element of $F = \hat{r}^k_{i+1} - \hat{r}^k_{i+1}$.

ZOH = zero order hold.

PID_i = proportional integral derivative controller on the ith row and j^{th} column.

G_{ij} = element of process transfer matrix in the ith row and j^{th} column. The transfer function between i^{th} output and j^{th} input.

DP/I = differential pressure to current transducer.

I/P = current to pressure transducer.

P/I = pressure to current transducer.

BP = body pressure.

VR = vent valve opening.

CA = control action called for by the steam flow primary controller.

DA = decoupling action called for as input to the secondary steam flow controller.

TABLE I

COEFFICIENT MATRICES IN DISCRETE MODEL FOR THE LABORATORY AUTOCLAVE

Row	Matrix A Column 1	2
1	0.989	0.006 - 0.00136*BP
2	0.204	0.775

Row	Matrix B Column 1	2
1	0.004+0.00072*BP	-0.0113+0.0236*BP
2	0.0247	-0.193

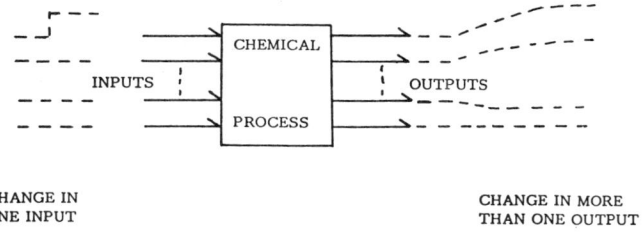

CHANGE IN ONE INPUT   **CHANGE IN MORE THAN ONE OUTPUT**

FIG. 1. INTERACTION IN MULTIVARIABLE PROCESSES

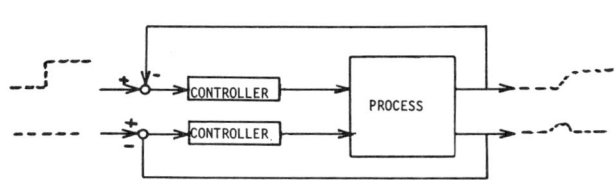

Change in one setpoint causes disturbance in second loop.
Controller transfer equation = $K[1 + \frac{1}{T_r s} + T_d s] E(s)$.

FIGURE 2. CONVENTIONAL CONTROL STRATEGIES APPLIED TO MULTIVARIABLE PROCESSES.

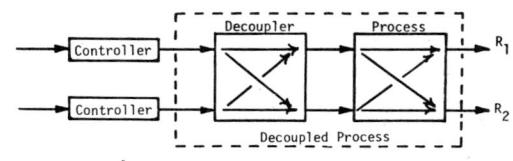

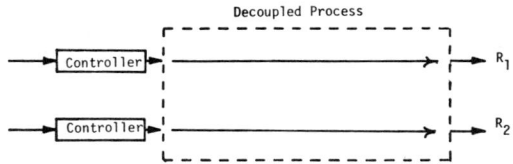

FIGURE 3. ILLUSTRATION OF DECOUPLED PROCESS

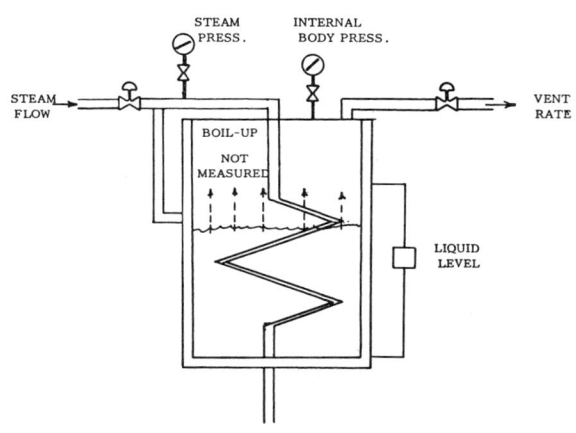

FIGURE 4. CRITICAL PROCESS VARIABLES.

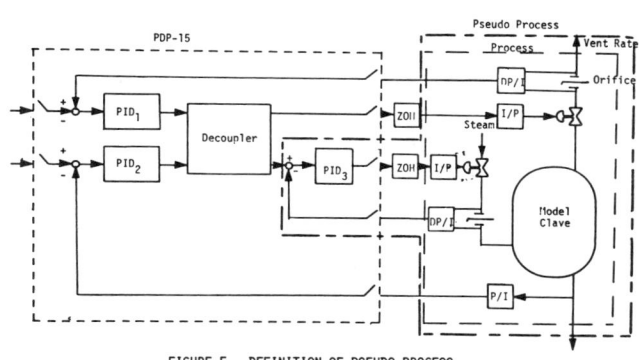

FIGURE 5. DEFINITION OF PSEUDO PROCESS.

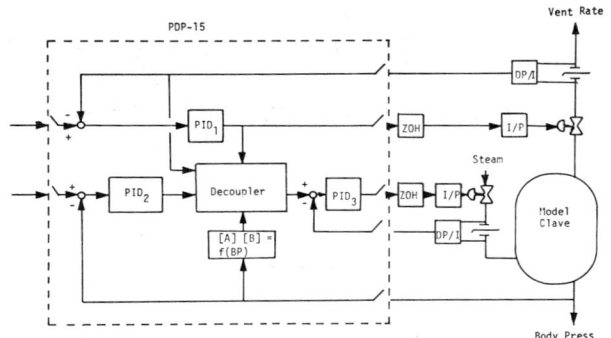

FIGURE 6. FINAL MULTIVARIABLE CONTROL STRATEGY.

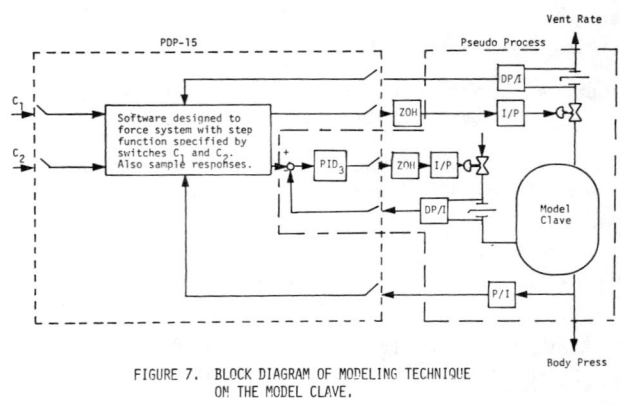

FIGURE 7. BLOCK DIAGRAM OF MODELING TECHNIQUE ON THE MODEL CLAVE.

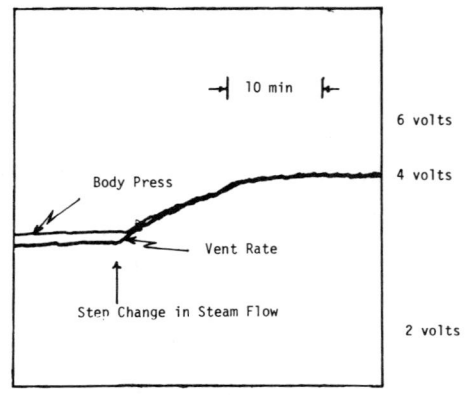

FIGURE 8. STEP RESPONSE OF MODEL AUTOCLAVE TO CHANGE IN STEAM FLOW.

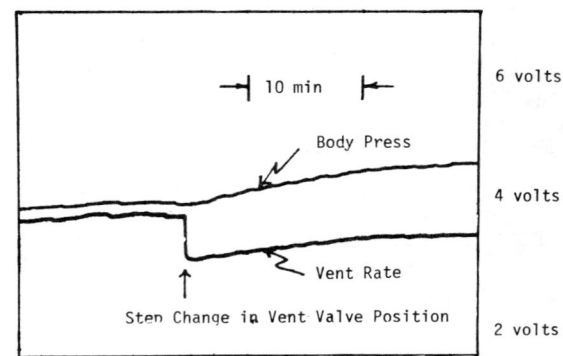

FIGURE 9. STEP RESPONSE OF MODEL AUTOCLAVE TO CHANGE IN VENT VALVE POSITION.

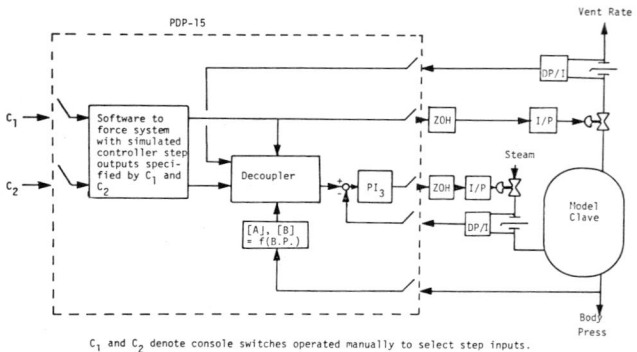

FIGURE 10. BLOCK DIAGRAM FOR OPEN LOOP DECOUPLER TEST ON THE MODEL CLAVE.

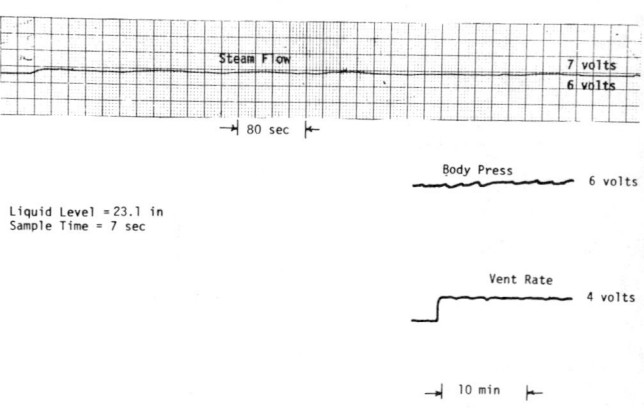

FIGURE 11. OPEN LOOP DECOUPLER FOR MODEL AUTOCLAVE, CASE I.

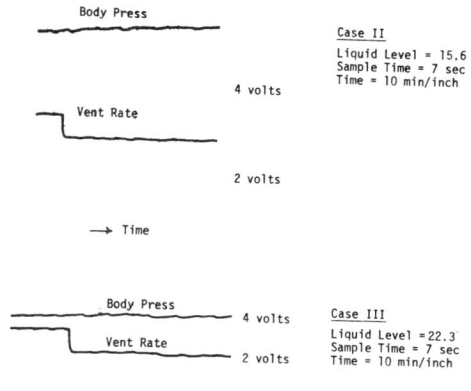

FIGURE 12. OPEN LOOP DECOUPLER FOR MODEL AUTOCLAVE CASES II AND III.

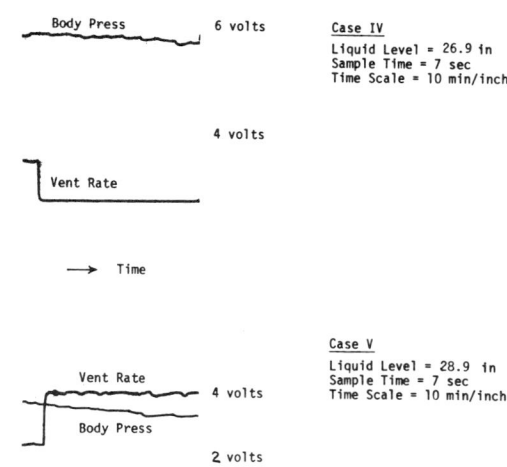

FIGURE 13. OPEN LOOP DECOUPLER FOR MODEL AUTOCLAVE, CASES IV AND V.

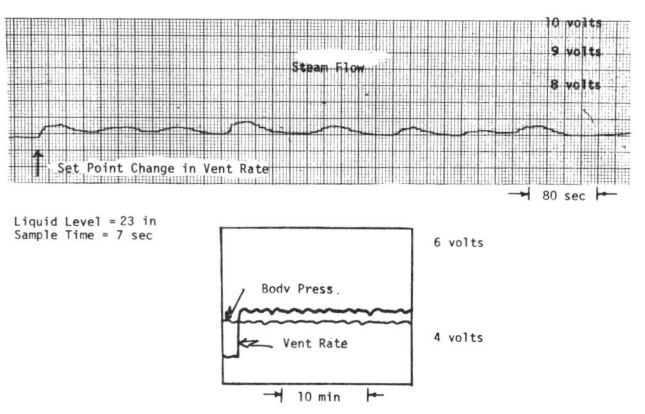

FIGURE 14. MULTIVARIABLE CONTROL OF MODEL AUTOCLAVE, CASE I.

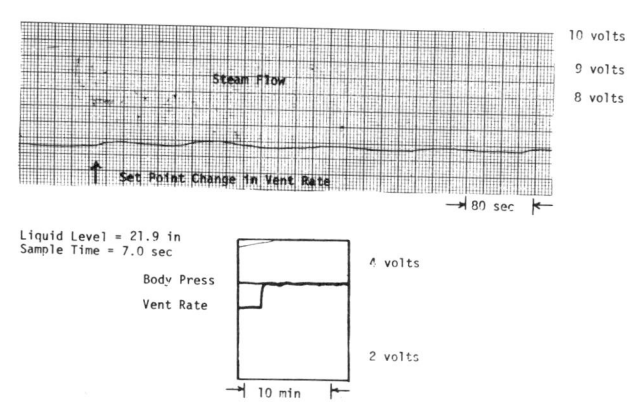

FIGURE 15. MULTIVARIABLE CONTROL OF MODEL AUTOCLAVE, CASE II.

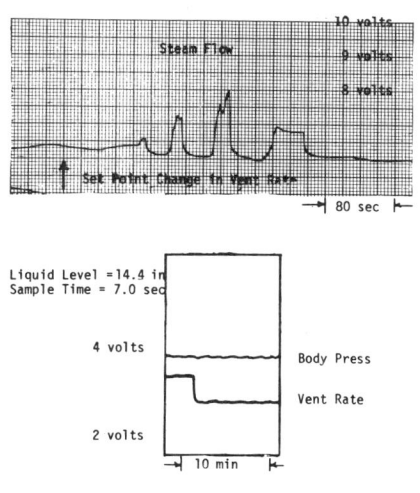

FIGURE 16. MULTIVARIABLE CONTROL OF MODEL AUTOCLAVE, CASE III.

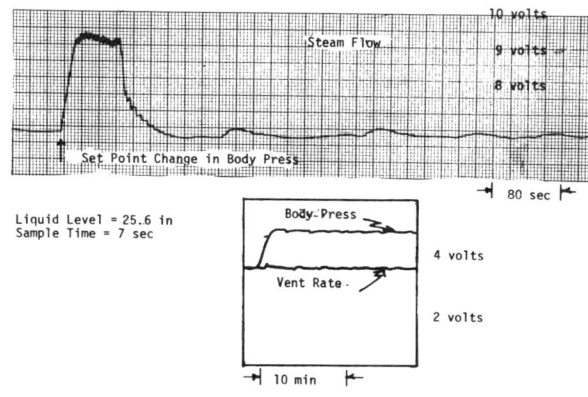

FIGURE 17. MULTIVARIABLE CONTROL OF MODEL AUTOCLAVE, CASE IV.

EVALUATION OF CONTROLLERS FOR DEAD-TIME PROCESSES

Charles W. Ross
Leeds & Northrup Company
North Wales, Pennsylvania

ABSTRACT

The paper discusses some model control schemes and evaluates them relative to a conventional two-mode controller from the design and application points of view. The evaluation factors include cost, varying process gains and dynamics, model mismatch, performance relative to noise and step disturbances, and tuning difficulties. The results of simulation studies are summarized for the control of processes which have a wide variety of combinations of dead time and first order lags. Rules of thumb may be obtained from curves in the paper for tuning the conventional controller and estimating its performance for random noise and step disturbances. Finally, the paper describes a simple but effective way to adapt the tuning and filtering for a conventional controller to processes which have time-varying dynamics.

INTRODUCTION

The dynamic characteristics of many industrial processes include dead times. Some examples are:

Heat flow in distributed parameter transportation systems.

Transmission over hydraulic and pneumatic lines.

Material transportation, e.g., conveyors, steel rolling, calendering, paper manufacturing.

Analyzers, e.g., chromatograph, X-ray, mass spectrophotometer, oxygen analyzer, etc.

A number of control schemes have been suggested to deal with the dead time control problem. Four schemes which have perhaps received the widest attention are described below. Digital techniques which employ a number of weighted delayed feedbacks will not be discussed here because it is difficult to generalize their form and tuning for various process characteristics. In general, the overall performance of these controllers appears to be inferior to the conventional controller.

Conventional Two-Mode Control

The test arrangement for the study of a controlled system with a conventional proportional-plus-integral controller is shown in Figure 1. No rate action is included in the controller because we assume there is a significant amount of dead time and noise in the process. The process characteristics to be studied are made up of various combinations of dead time T_D and first order lags with time constants of T_1 and T_2.

Model Feedback Control

A simple form of a model feedback controller proposed by Reswick[1] is shown in Figure 2. This controller uses a model of the process G_p' in the positive feedback path in addition to the negative proportional feedback. A requirement to prevent steady-state error with model control is that $G_p'(0) = 1$, where $G_p'(0)$ is the process model evaluated at zero frequency. That is the loop gain in the positive feedback must be unity. The negative feedback controller gain K_c is then adjusted to obtain the optimum response.

Dead Time Control Scheme

The Smith[2] control scheme shown in Figure 3 assumes that a process can be characteristized by a dead time T_D in cascade with a minimum-phase function $G_1(s)$. In this application it will be assumed that G_1 consists of first order lags T_1 and T_2. Again in order to prevent steady-state error the gain in the dead time part of the model must be unity. In essence the feedback from control action takes place immediately through the model $G_1'(s)$. The part coming through the model dead time will be cancelled by the response through the process. Therefore, if the model matches the process, the conventional controller G_c may be tuned to give optimum response for the minimum-phase part of the process G_1. Reswick and Smith's schemes are essentially the same for pure dead time processes.

Wait-and-See Control

Wood[3] has suggested an interrupter type dead time control. This scheme basically consists of a relay which disables the controller for a period equal to the dead time after action is taken. Although this control scheme, including other forms of wait and see control, is rather simple to implement and use, its performance is poorer than those described above. The performance is further degraded for processes which also have appreciable lags and noise.

TUNING DIFFICULTIES

Traditionally, dead time systems have been termed "hard to control." Actually, the tuning of a controller for a dead time process may be simpler than for other types of processes. Consider, for example, the tuning of a conventional controller for a dead time process. If the controller gain is set at 0.3 (PB = 333%) and the reset rate is $2.5/T_D$, the response to a step disturbance is as shown in Figure 4. It should be noted that no control action takes place before one dead time T_D. At this time the disturbance is reduced by 30%. During the next dead time period the disturbance is reduced linearly to -5% of the step size. At $2T_D$ the feedback is such as to bring the disturbance to zero. During the next period the response is slightly oscillatory but the overshoot is limited to 5%.

The ideal response which is possible for a perfectly tuned wide bandwidth controller (model control) is a rectangular pulse having a duration equal to the dead time as shown in Figure 4. It is interesting and important to note that the same response is obtained regardless of where the dead time is in the process or in the feedback loop. Also for a pure dead time process the same controlled response is observed for a disturbance at the input or output of the process.

Many analog controllers can only adjust the reset rate by factors of two. A doubling of the reset results in considerable oscillations as shown in Figure 4. The response can be brought back to near optimum by doubling the proportional band (PB). The effect of gain and reset changes for a conventional controller on a pure dead time process are shown in Figure 5.

Suggested two-mode controller settings are given in References 4, 5 and 6 for various combinations of dead time and lag processes. These settings are useful not only for field tuning but also for the design of adaptive control schemes. These references also take into account the effect of sampled data on controller tunings and control performance when direct digital control DDC is used.

The effect of gain changes in the model control is easily seen by tracing the ideal square responses through the control and process loops. For example, if $K_c = 2$, there will be a continuous cycling with a period of $2T_D$. The system will be unstable for $K_c > 2$. For $K_c < 1$ the response approaches the setpoint in a series of steps occurring at every dead time.

For pure dead time systems any mismatch between the controller model and the process can be disastrous to the performance of wide bandwidth controllers such as the model control. This mismatch results in the feedback being in and out of phase and can lead to a serious instability problem. Of course, filtering can be applied but this results in decreased closed loop performance of the controller. It is important to note here that additional lags added to the model only will make the situation worse. This is due to the fact that lags in the model feedback has the effect of adding rate in the forward path. Finally, if the model control is to be executed digitally for $T_D \gg T_P$, great care must be taken in its design with regard to sampling rate, coordinating the sampling, adding filtering and chosing controller tunings. When all of this is done the overall performance may be no better than that of a conventional controller. The conventional controller has a very important advantage here because it tends to be unaffected by small mismatches between its tuning, sampling rate and the process dynamics.

RESPONSE PERFORMANCE FOR STEP DISTURBANCES

The relative performance of the various control schemes will depend to a considerable extent on the process, the criteria chosen for evaluation and the manner of testing, e.g., the nature of the disturbance and where it is applied in the system. The manner of testing discussed under this section will be the same as that discussed in Reference 4. The process consists of a dead time and two simple lags ($T_1 = 3T_2$). The tuning is for a minimum ITAE and 5% overshoot. The disturbance is a step function applied at the process output.

(a) $T_1 = T_2 = 0$:

One extreme case is a pure dead time process. For this case the wide bandwidth model controllers (properly tuned) will reduce error to zero after a period of one dead time. This is the best performance possible with a feedback controller. The percent degradation of the conventional controller to this ultimate performance is

$\%t_s$ = 75% - Settling time - The conventional controller requires 75% longer to reduce a disturbance to within 10% of its steady-state value.

%ITAE = 100% - Integral of time times the abolute error.

%IAE = 30% - Integral of the absolute error.

From the practical point of view the IAE criterion may be the more indicative with regard to most process performance requirements. It should be pointed out that the conventional controller would show even better relative performance for the ISE (integral squared error) criteria. The relative superiority of the wide bandwidth controller will be reduced for other types of disturbance, e.g., ramp disturbances.

(b) $T_P \geq T_D$

The Smith circuit outperforms the other control schemes when the process has dead time and lags. This is due to the controller in the circuit which is tuned for optimum response for the lags in the process.

The relative improvement of Smith's circuit over the conventional controller may be evaluated for the settling time criteria from Equations (3) through (6) in Reference 4. The settling time t_s for Smith's circuit for output step disturbance and $T_P = T_1 + T_1/3$ is

$$t_s \approx 0.7 T_P + T_D \qquad (1)$$

Note: $t_s \approx 1.0 T_P + T_D$ for $T_1 = T_2$

The relative improvement in settling time compared to the conventional two-mode controller response t_0 is given for disturbance at the output as

$$t_s/t_0 \approx \frac{0.7 + T_D/T_P}{0.7 + 2.38\, T_D/T_P} \text{ for, } T_D/T_P < 1.0 \qquad (2)$$

$$t_s/t_0 \approx \frac{0.7 + T_D/T_P}{1.33 + 1.75\, T_D/T_P} \text{ for, } T_D/T_P > 1.0 \qquad (3)$$

For a step disturbance applied at the input of the process,

$$t_s/t_0 \approx \frac{1.4 + T_D/T_P}{1.4 + 2.38\, T_D/T_P} \text{ for, } T_D/T_P < 1.0 \qquad (4)$$

$$t_s/t_0 \approx \frac{1.4 + T_D/T_P}{2.03 + 1.75\, T_D/T_P} \text{ for, } T_D/T_P > 1.0 \qquad (5)$$

Equations (2) through (5) are plotted in Figure 6 versus the ratio T_D/T_P. These curves indicate a considerable improvement for the Smith circuit. However, as indicated earlier, settling time is probably too severe a criteria and IAE may be a better index of performance which shows about one-half the improvement. Also the improvement will be reduced if the disturbance is not a step function. The improvement will be further reduced (especially for cases approaching pure dead time) if the model is not correct. From the step performance point of view, the Smith circuit appears particularly attractive for cases where $T_P \approx T_D$ since Padé and sampled data circuits may be used more effectively for modeling the dead time, and the noise and stability problems are reduced.

CONTROLLABILITY OF STATISTICAL DISTURBANCES

Some guiding principles for determining in advance to what extent statistical (noise) disturbances may be controlled in a given system would be helpful for deciding how the disturbance should be conditioned prior to its transmission to the controller. Such decisions require trade-offs between filtering to reduce the effect of noise and the degradation of control, e.g., to a step disturbance.

In practice the criteria which may be used for deciding how the disturbances should be controlled may depend on a number of factors which are weighted differently for various applications. A starting point for many applications is to determine the answers to the following questions:

1) For a given controlled system, to what extent can a given disturbance be controlled?
2) How much control activity is required to accomplish this degree of controllability?

A simulation study was carried out to determine the answers to these questions on a number of system configurations. The general test set-up is illustrated in Figure 1. The disturbance was random noise shaped by a first order lag having a time constant τ_n. The proportional and reset action controller was adjusted to give a 5% overshoot for control of a step disturbance. The following system characteristics were evaluated:

1) Pure dead time, T_D : $T_{PS} = T_D$
   where, T_{PS} is the process storage time.
2) Dead time plus two first order lags,
   $T_1 = 3T_2$: $T_{PS} = T_D + T_1 + T_2$
3) Two first order lags,
   $T_1 = T_2$: $T_{PS} = T_1 + T_2$
4) Two first order lags,
   $T_1 = 3T_2$: $T_{PS} = T_1 + T_2$

Under each test condition measurements were made of the mean-square values for

1) $\int N^2 dt$ - noise disturbance
2) $\int \epsilon^2 dt$ - error due to disturbance
3) $\int M^2 dt$ - manipulated variable

The test results have been summarized in Figures 7 and 8. The results have been made more generally applicable by normalizing the noise dynamic characteristics to those of the system, i.e., τ_n/T_{PS}. The error and manipulated variables have been normalized to the disturbance measurement. Values for τ_n may be estimated in the field from operating records as discussed in References 6 and 7.

It should be noted from Figure 7 that the noise seen at the output of the system is actually worse due to control action for high frequency disturbances, i.e., for smaller values of τ_n/T_{PS}. Also, it is seen that disturbances are less controllable in systems having dead time. On the other hand, comparing Figures 7 and 8, the controller activity for high frequency disturbances is much less in dead time control than for systems having lags. This is due to the much smaller loop gain used in dead time control.

Model controllers exhibit very poor performance for noisy processes because of their high gains and bandwidth relative to a conventional controller. Consider, for example, a pure dead time process optimumly tuned for a step response, i.e., $K_c = 1$. For high bandwidth noise, $\tau_n/T_D \ll 1$, the mean square error seen at the output of the process is increased by a factor of two compared to a 10% increase for a conventional controller. The normalized mean square movement of the manipulated variable is 2 for the model control compared to only 0.12 for the two-mode controller. The situation may be worse for other values of τ_n/T_D. No effective control can be taken on disturbances with durations less than $2T_D$.

GENERAL EVALUATION OF CONTROL SCHEMES

When evaluating a particular control scheme a number of factors must be considered. A number of these factors in their probable order of importance are listed below.

1) Cost
2) Noise - uncontrollable disturbances
3) Variability of the dead time, ΔT_D
4) Performance vs. ratio of dead time to lag time constant, T_D/T_P
5) Variability of process and controller gains, ΔK_c
6) Controller tuning difficulties

A methodology for proceeding with an overall evaluation is illustrated in Table I.

TABLE I

AN EXAMPLE OF EVALUATION FOR
RELATIVE PERFORMANCE OF VARIOUS CONTROL SCHEMES

Control Approach	Cost	Noise	ΔT_D	Step Performance $T_P < T_D$	Step Performance $T_P > T_D$	ΔK_c	Diff. Tun.	Wt. Total
Conv. PI	1	1	2	4	1-2	1	1	28
Smith	3	2-3	4	1	1	3	4-5	48
Reswick	2	2-3	4	1	2-3	2	4	44
Wood	1-2	4	3	3-4	3-4	2	2	51
Weighting	5	4	3	2	2	1	1	

The evaluations in the Table are in relative terms; the smaller the number the better the performance. For a particular control problem the various factors might be weighted differently. For example, if step performance is quite critical and outweighs the other factors, the Smith circuit would probably be chosen. The weighting factors used in the Table are quite arbitrary but are based on discussions with a number of people and are intended to cover the more frequently encountered cases. For general use the Table indicates that conventional two-mode controller is preferred. This is because of the wide use of the conventional control and its robust control characteristics.

In many cases it is not possible, in advance, to place numerical values on the criteria for evaluating a particular controller. The purpose of the following general discussions of each factor in Table I is to aid in the choice of a controller for a particular application.

Cost - For analog execution the model controllers will generally cost considerably more than conventional controllers. The model controller not only requires more hardware but the number required is small and they are not generally commercially available. If execution is done digitally in a larger control computer the differential in hardware cost is small but there will be an increment in cost for the model control for its design and application.

Noise - If the process has considerable noise the conventional controller will generally be preferred. This is particularly true for pure dead time processes. Any heavy filtering and/or reduced gain with model control will degrade its performance for other disturbances such as step changes.

Variable dynamics - Changes in the dynamics of the process, especially T_D, will degrade the performance of the model control more rapidly than for conventional control. Of course this difficulty may be overcome if the dead-time model is made to vary adaptively with the dead time of the process.

Performance for step changes - The step-response performance is about the same for both controllers as the ratio of dead time to lag time constants approach zero, i.e., $T_D/T_P \ll 1$. For most cases where dead time is appreciable the model control should give better performance. The relative increase in performance shown in Figure 6 may be further increased for the Smith control in the mid-range if rate action can be used. The performance, however, will be decreased for $T_D/T_P \ll 1$ due to the difficulties of realizing a perfect model and for slow varying disturbances such as ramps.

Variable gains - The degradation of control performance is somewhat greater for model control when there is a change in controller or process gain. Also care must be taken in the model control path to insure a gain of unity to prevent steady state errors.

Tuning difficulties - As was discussed earlier tuning will generally be much more difficult for model control, especially for $T_D/T_P \gg 1$. Although tuning was given only a weighting of one in Table I, it may deserve a higher weighting from the application point of view.

ADAPTIVE DEAD-TIME CONTROLLER

Any of the controllers discussed above may be made adaptive by adjusting the controller (and/or model) in conformance with the system gain and dynamics if they are known and are measurable. This may become somewhat involved if controller tunings are varied in a nonlinear manner to maintain optimum settings over a wide-range of T_D/T_P. Additionally, for the model control it is generally quite difficult to obtain a suitable dead-time model which can adequately cover a wide range either by analog or digital methods. This section will deal with a simplified approach for making a conventional controller adaptive for wide ranges of dead time T_D and lags T_P.

Studies of two-mode (PI) controller settings for processes which consist predominately of dead time have indicated the following: (4), (5), (6)

1) The controller gain is essentially independent of the process dynamics and depends only on the system loop gain.
2) The reset setting decreases as the dead time and/or the process lags increase.

As an example, an essentially optimum tuning which results in about 5% overshoot is:

Total loop gain (process & controller = 0.3)

$$\text{Reset rate} \approx \frac{2.5}{T_P + T_D} \qquad (6)$$

where: T_P = the sum of the process lag time constants
T_D = process dead time
$T_D/T_P > 1$

The optimum tuning is changed, however, as the value of T_P approaches and/or becomes larger than T_D. The optimum controller settings for a unity gain process with two lags ($T_1 = 3T_2$) are illustrated in Figure 9. The resulting system responses for these settings are shown in Figure 10 for various values T_P/T_D. It should be noted from Figure 9 that the above control settings give the optimum responses for cases of $T_P/T_P > 2$. The breakpoint will be somewhat larger than 2 in a single lag system and occurs at a smaller value as T_2 approaches the value of T_1.

Assume now that the controller settings are maintained adaptively at the optimum values for essentially dead-time systems (Equation 6) for all values of T_D/T_P. The resulting system responses for a step disturbance will be as shown in Figure 11 and the penalty paid in the IAE criterion is given in Figure 9. It should be noted from these figures that the loss in performance is essentially negligible for cases for $T_D/T_P > 1$. Even for cases of $T_D/T_P < 1$ the system is stable and the overshoot remains less than 5%. For any systems that vary over such a wide dynamic range as that illustrated, the loss in performance would probably be acceptable for general applications.

A comparison of Figures 10, 11 and 12 indicate the vast improvement of an adaptive controller over a nonadaptive conventional PI controller. Figure 12 illustrates a conventional controller tuning for the case of $T_P/T_D = 0.5$. The dead time was then reduced by only a factor of two and the response became extremely sluggish. When the dead time was increased by only a factor of two the system response became oscillatory. This response should be compared to Figures 10 and 11 where the essentially optimum response is maintained.

The system response can be improved for varying dead time if the controller gain is varied inversely as the dead time. This is illustrated in Figure 13. The response, however, deviates considerably from the optimum and near optimum tunings shown in Figures 10 and 11. The response in Figure 13 could be improved somewhat if a constant plus a variable gain had been used. In all of the adaptive schemes the controller gain should be changed to compensate for the process gain if it varies as a function of the dead time or other process parameters.

Most dead-time processes have uncontrollable noise at their output. Any attempt to control this noise only results in an increase in the output noise and unnecessary wear on the control actuators, valves, etc. It may therefore be desirable to use an adaptive filter to remove as much of this noise as possible for a given degree of performance degradation. For the maximum noise reduction consistent with a given performance degradation the filter time constant may be increased as the process dead time increases. It should also be noted that an adaptive filter is highly desirable for an adaptive model control. This is especially true for pure dead times processes to reduce to model mismatch problem.

A typical manner in which the adaptive filter may be used is illustrated by the following example. One may judiciously choose to place filtering into a system such that the closed loop step response is degraded by a fixed amount. The required filtering is approximately determined from Reference 6 and Equations 2-5 in Reference 4 for the settling time criteria. Once the filter is adjusted it will automatically track the process variable T_D and give the same relative degradation for all values of systems parameters. It should be noted that the value of the filter time constant must be included when determining the reset value and is considered as part of the variable dynamics. The controlled system will then be adaptive and optimally tuned for all values of the variable dead time.

CONCLUSIONS

The conventional two-mode controller will generally be preferred over other control schemes for processes having considerable dead time. This controller is more robust for dealing with the wide variety of control considerations. Its performance, based on experience, is highly predictable. It is easy to implement, tune, and adapt to varying process characteristics. The Smith dead-time controller is still considered to be the nearest thing to an optimum controller for certain types of disturbances and process conditions but one should evaluate whether or not the added complexity is worth the effort.

NOMENCLATURE

- ϵ — control error
- G'_1 — model of first order lags in process
- G_c — controller characteristics
- G_p — dynamic characteristics of a process
- G'_p — model of dynamic characteristics
- $\int N^2 dt$ — mean-square noise at output of process
- K_c — controller gain
- $PB = 100/K_c$ — proportional band in %
- $R = 1/T_i$ — repeats per unit time
- τ_n — time constant of noise shaping filter
- t_0 — settling time for conventional controller
- t_s — settling time for Smith's model controller
- T_D — dead time
- T_i — integral time of controller
- T_1, T_2 — first order lag time constants
- $T_P = T_1 + T_2$ — sum of lag time constants
- $T_{PS} = T_P + T_D$ — storage time of process dynamics

REFERENCES

(1) Reswick, J. B., "Disturbance Response Feedback - A New Concept," TRANSACTIONS OF ASME, Vol. 78 p.153, January 1956.
(2) Smith, Otto, J. M., "A Controller to Overcome Dead Time," ISA Journal 6 (2), p.28, February 1959.
(3) Wood, John, "Controlling a Pure-Delay Plant," Instruments and Automation, Vol. 30, p.1720, September 1957.
(4) Ross, C. W., "Control of a Process with Analyzer in the Feedback Loop," ISA TRANSACTIONS, Vol. 2, Issue No. 1, January 1963.
(5) Fertik, H. A. and Ross, C. W., "Direct Digital Control Algorithm with Antiwindup Feature," ISA TRANSACTIONS, Vol. 6, No. 4, pp.317-328, 1967.
(6) Fertik, H. A., "Tuning Controllers for Noisy Processes," ISA TRANSACTIONS, Vol. 14, No. 4, 1975.
(7) Ross, C. W. et al, "Experimentally Determined Properties of Exponentially Correlated Gaussian Noise," ISA TRANSACTIONS, Vol. 13, No. 2, pp.172-181, 1974.

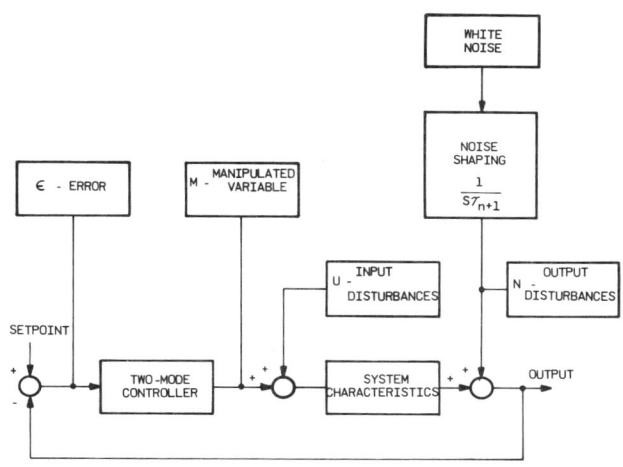

Figure 1 - Test Arrangement for the Study of a Conventional Controller

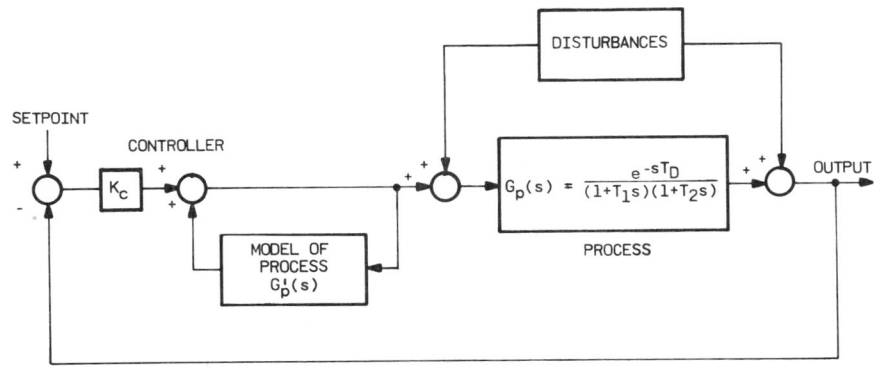

Figure 2 - Reswick's Model Feedback Controller

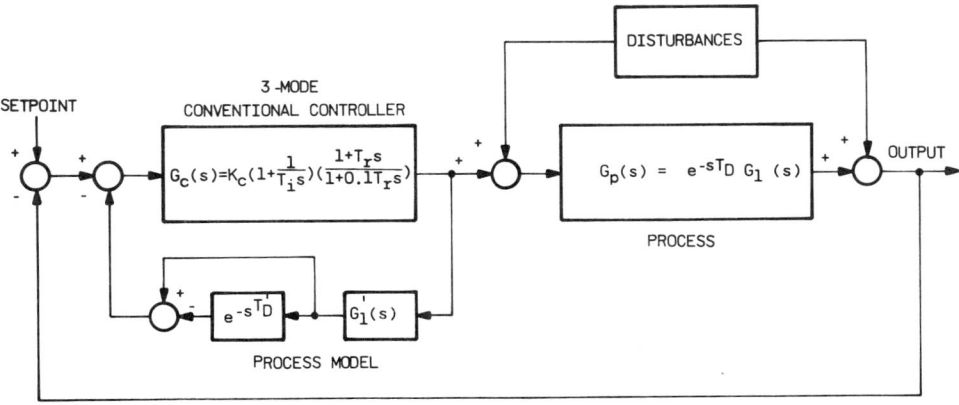

Figure 3 - Smith's Dead-Time Controller

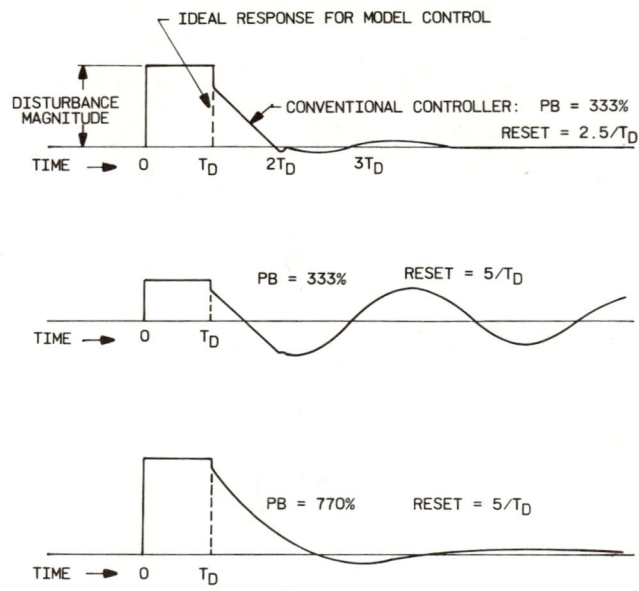

Figure 4 - Control Response for a Pure Dead Time Process

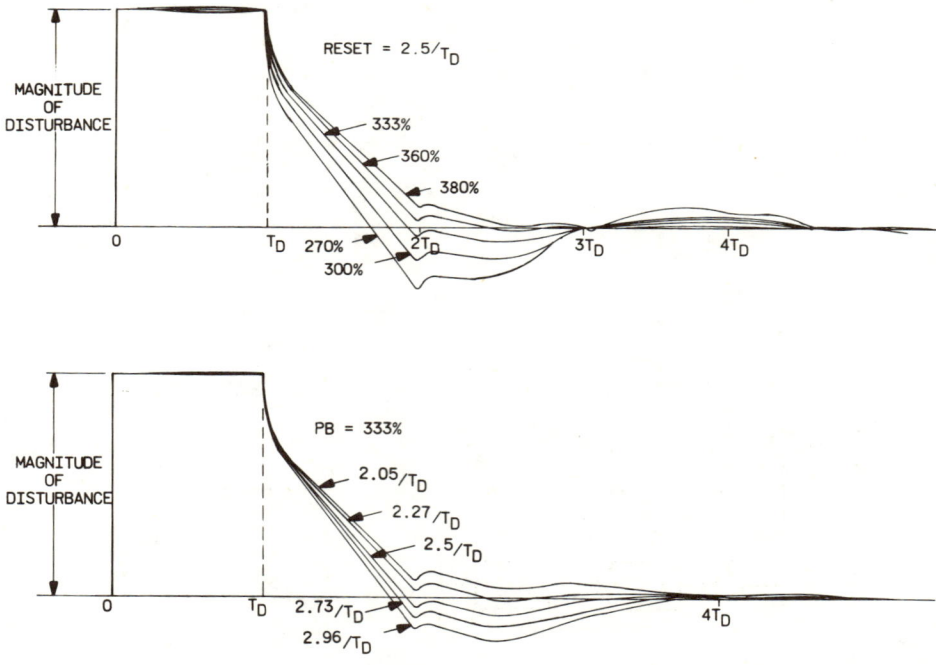

Figure 5 - Effect of Gain and Reset Changes for a Conventional Controller on a Pure Dead Time Process

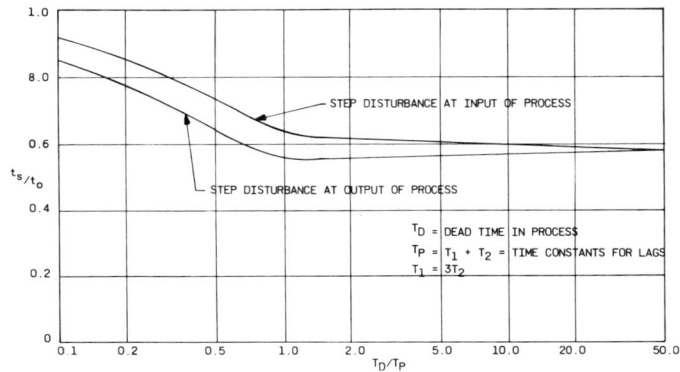

Figure 6 - Ratio of Settling Times: Smith's Method to Conventional Controller

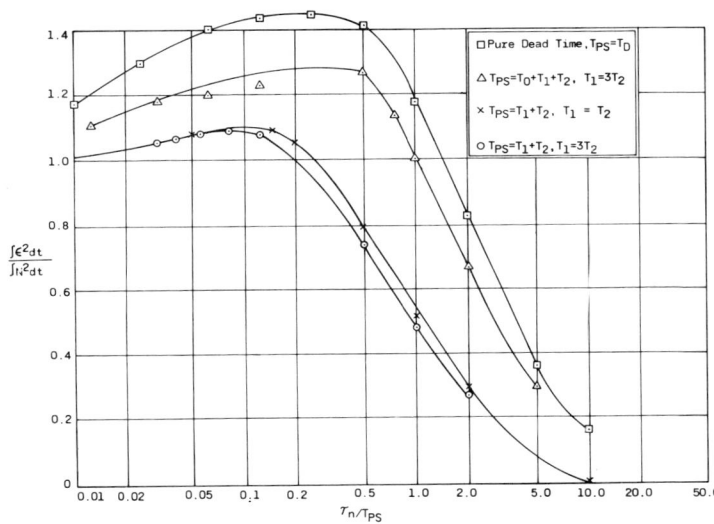

Figure 7 - Controllability of Statistical Disturbances with Conventional Two-Mode Controller

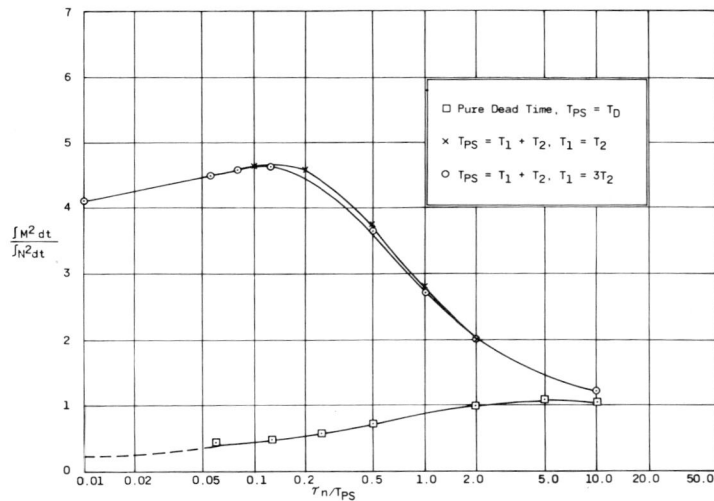

Figure 8 - Activity of Manipulated Variable for Statistical Disturbances with Conventional Two-Mode Controller

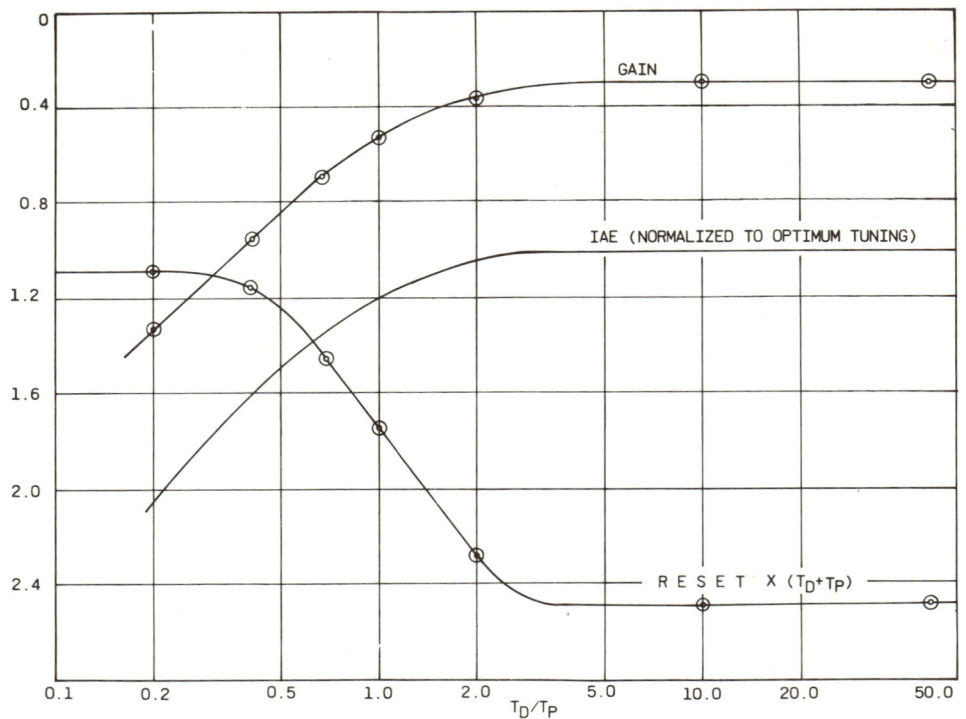

Figure 9 - Optimum Conventional Controller Settings, IAE for Approximate Tuning
(Equation 6)

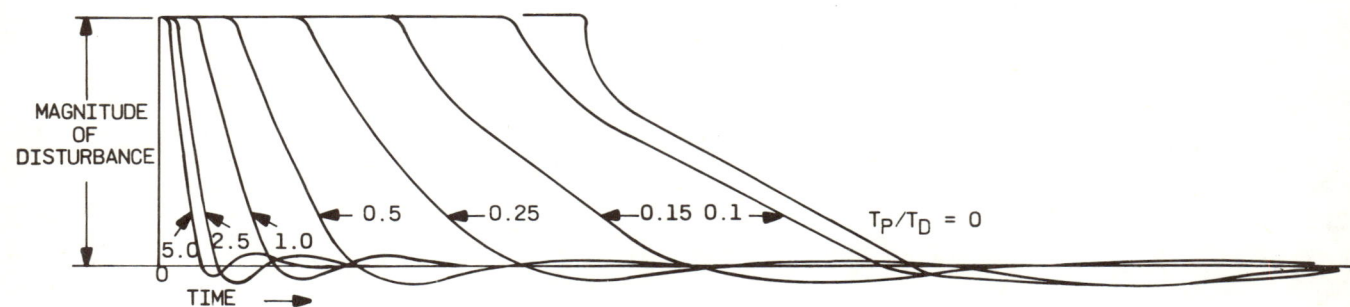

Figure 10 - Optimum Responses of Systems with PI Control

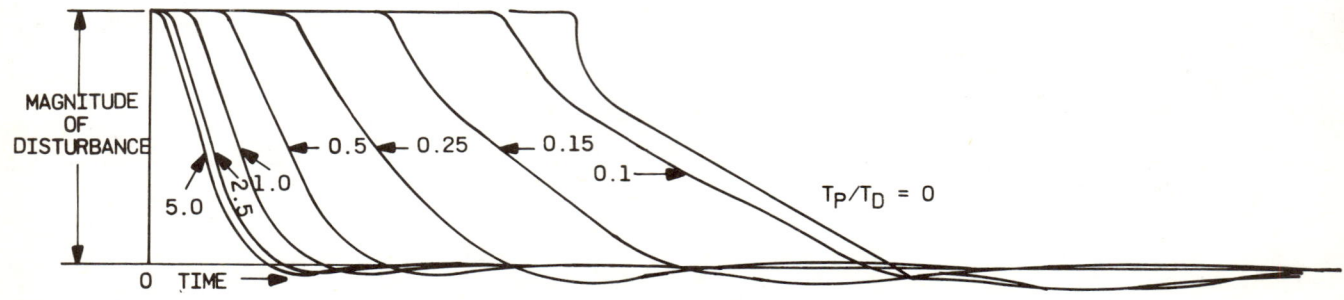

Figure 11 - Responses of Systems Using Fixed Gain and Variable Reset

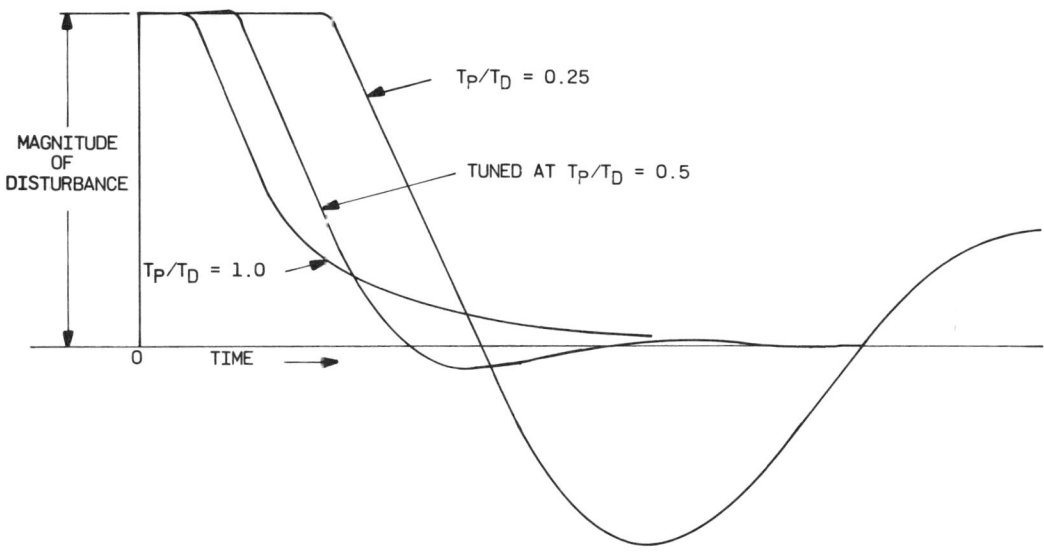

Figure 12 - Effect of Varying Dead Time on System Response - PI Controller with Fixed Settings

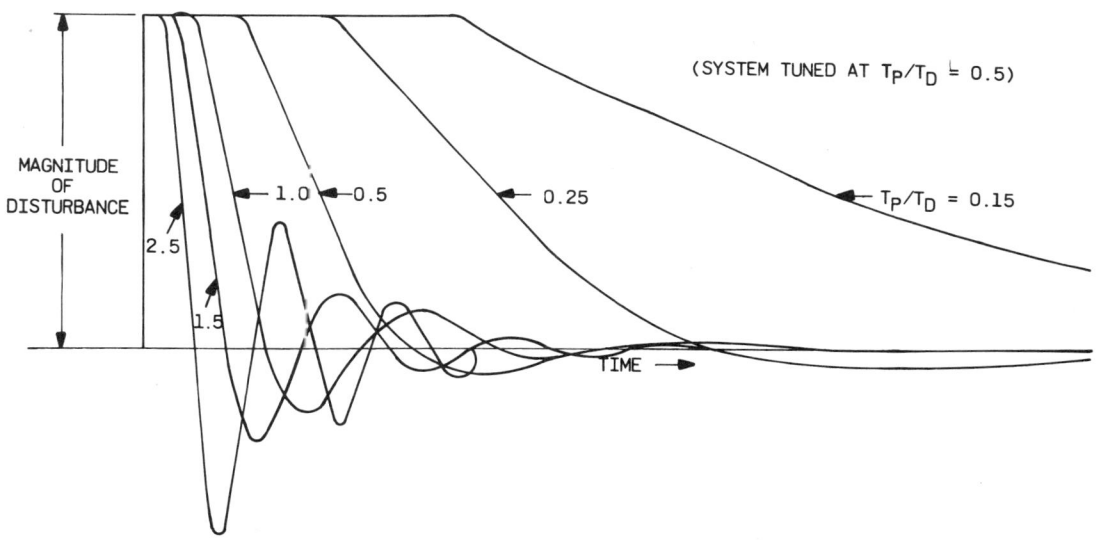

Figure 13 - System Response for Gain Varied Inversely as Dead Time, Reset Fixed

©ISA, 1977
ISBN 87664-363-2

DIGITAL COMPUTER CONTROL
OF AMMONIA PLANTS

James S. Gruneisen
Applications Engineer
Fisher Controls Company
Marshalltown, Iowa

ABSTRACT

Computer control of ammonia plants has been successfully achieved in twenty plants throughout the U.S. The control system was designed to assist, not replace, the operator. The speed and mathematical capabilities of the computer system allow successful control of synthesis loop H/N ratio, high pressure purge, furnace temperature, reform steam/gas ratio, and converter bed temperatures. This successful control results in production rate improvements, more efficient operation and fewer shutdowns. Now with proven results in many installations of 1/2 to 3% production increases, a computer can be justified with less than a year payout period.

INTRODUCTION

Experimentation with computer control of ammonia plants began in 1959 with the installation of a Ramo Woolridge RW300 digital computer. This first digital control system was installed in Monsanto's Chemico Ammonia Plant at Luling, Louisiana. Since that initial installation, approximately 25 ammonia plants have digital control systems on line or on order. There are many reasons ammonia plants are prime candidates for computer control, but the underlying justification is that with more sophisticated control of certain process variables and additional operator information, increased production and/or efficiency can be attained. This fact has been verified by computer simulations and by users that have installed digital systems.

The initial system included control of two trains. The process variables under control in each train included process gas, two reformer fuel flows, reform steam, secondary air, and CO converter steam. The primary purpose of the control scheme was to load the reciprocating compressors for maximum production in face of varying ambient conditions. A secondary task was to maintain synthesis loop H/N ratio. A significant increase in production was attained due to proper control of the H/N ratio and compressor loading. Monsanto experienced an increase in production of 15 TPD.

Based upon the success at Luling, a digital control system was implemented at El Dorado, Arkansas. This system operated successfully until the plant was dismantled. The system at Luling was expanded in 1965 to accommodate the additional programming requirements for control of a new 600 TPD Kellogg plant that was to be placed on line in 1967. This new control scheme on the Kellogg ammonia plant included control of loop pressure, synthesis H/N ratio, primary furnace temperatures and quench flows to the synthesis converter beds.

CURRENT INSTALLATIONS

The current installations can be grouped into two categories. The first category is a digital control scheme designed primarily for control of several loops within the plant. The second category consists of a digital control scheme for control and sophisticated operator information system. Basically the control only system would consist of the equipment outlined in Figure 1. Operator entries would be made through the hard copy unit. A short production efficiency log would be printed on an hourly, shift and daily basis. The production efficiency log would consist of process gas flow, fuel flow, ammonia flow, purge flow, and other significant efficiency indication flows or averages.

The control scheme in both systems performs the same tasks, however, the second system includes a sophisticated operator information system. The equipment for this system is outlined in Figure 2. The operator information consists of two program modules. One module handles all operator entries using an interactive conversational method and a CRT. The operator begins a sequence on the CRT by typing a keyword. The system replies with questions pertaining to the task requested. Once the system has asked the appropriate questions and the operator has made reasonable replies, the task will be completed. To short cut the question and answer period, the operator may enter the first three letters of the keyword and the answers to the questions separated by commas on the same line. A sequence of entries using this routine is outlined in Figure 3. There are three distinct advantages to this system - operator training, system security and system flexibility. Operators can be trained in much shorter time since they must only learn the answers to a number of questions. There is no memorization involved, which means operator acceptance is gained more rapidly. System security is enhanced since all entries to the system are checked for validity. The system will only accept keywords or answers to previously asked questions. By using a CRT and alphanumeric keyboard, an engineer can easily add or modify keywords or questions sequences. This adds flexibility to the system since no changes need be made to keep the operator entry system up to date.

The second operator information module includes a display formatting program. This program allows the operator or engineer to define a display on the CRT screen. The sequence begins by simply typing the alphanumeric display on the CRT screen. Once complete, the display is given a number and transmitted to the computer. The final phase involves a definition of the values to be printed as a part of the display. The displays can be combined into a log or used as a continuous display on the CRT screen. A hard copy can be requested at any time. Any of the displays can be documented by a simple request. The documentation phase prints the definition of each number printed as a part of the log. New displays can be defined online or old displays modified to meet current requirements. This program gives the user an extremely flexible information system. It allows the engineer to define displays useful for system startup and later add displays for the operating personnel. A display can be defined for specific areas of the plant for a plant study. All of these displays can be built and stored in a matter of a few minutes to provide useful and pertinent information for the operator or engineer.

AREAS OF CONTROL

As mentioned earlier, the control portion of both systems consists of basically the same control scheme. Each control scheme is designed on an individual plant basis. This requires combining of various standard control modules and modules designed specifically for the plant into a digital control scheme. In design of the control scheme, the design must combine past experience with the needs of the plant. Several criteria are used to select the areas of the plant to control. One may begin the selection process by listing all process variables that affect efficiency and/or production in their relative order of sensitivity. The next step is to further narrow this list by selecting the process variables that cannot be adequately controlled by the operator and a conventional analog control scheme. The list can be further narrowed by selecting the top 20-50% of the variables as an initial system and saving the others for future expansion.

Based upon past experience, five loops can be recommended for all plants. In addition to these loops, control schemes should be added to satisfy the specific needs of the plant. The areas of control usually recommended are synthesis loop H/N ratio, synthesis loop purge, primary furnace temperature control, synthesis converter temperature control and reform steam/ gas ratio control. To attain measurable production and/or efficiency increases all the above loops should be digitally controlled. This can be understood by considering that ammonia plant is continuous and that poor control in the front end of the plant can actually prevent a digital system from attaining good control in the back end of the plant. Likewise, good control of the front end of the plant will not provide production and/or efficiency increases if the synthesis loop is not controlled properly.

In ammonia plants, the variable most difficult to control is H/N ratio at the inlet to the synthesis converter. This is due to the difficulty involved in controlling a variable with a time constant of fifty to one hundred minutes. The time constant is the time between changing air or gas and the time the H/N ratio comes to 67% of its new final value. In addition to the long time constant, there is also a five to ten minute dead time due to the transportation delay involved when an operator makes an air flow change and the time it is detected at the back end of the plant. In order to measure the H/N ratio, an on-line gas chromatograph must be utilized. The chromatograph is often difficult for the operator to utilize since data must be first scaled and then divided to attain the current ratio. The digital control scheme is designed to give very exact control of the H/N ratio. The normal standard is control within a band of ± 0.05 ratio units from the setpoint. In most plants this is an order of magnitude better than an operator can attain. This improved control will allow the process engineer to set the ratio at a constant value for maximum efficiency or production. The engineer should adjust loop H/N ratio setpoint for different plant operating conditions. In some systems, the set point is adjusted by a computer optimization program periodically. Obviously, a digital system is required for the optimization program, but a digital system is also required for the operator or the process engineer to manipulate the H/N set point intelligently. The digital system will present up-to-date information in engineering units so that process engineers can evaluate plant operation to make intelligent decisions concerning the proper H/N ratio to operate at. The H/N control program will aid in plant rate changes by properly manipulating the air or gas using feedforward techniques to maintain loop H/N ratio within $\pm.05$ ratio units.

The digital system includes a gas chromatograph interface program. This program provides for on-line automatic calibration of the gas chromatograph. The system counts peaks and checks each peak for a reasonable value before it is stored. If the sum of the components is not within 2% of 100%, the operator is warned of excessive error. All of these features enhance the gas chromatograph capabilities and make the gas chromatograph a more reliable piece of equipment.

Proper control of the purge valve will increase production. The control scheme maintains constant loop pressure. However, in plants having centrifugal synthesis gas compressors an optimization program varies the loop pressure set point to assure maximum throughput when the synthesis gas compressor becomes a bottleneck. Loop pressure control utilizing the digital system can be maintained within a band of ± 5 psi. Only a digital system and a special pressure transmitter will allow pressure within ± 5 psi of set point. Through a computer simulation of the synthesis loop and practical experience, it has been determined that the higher the pressure the loop can be operated at, the more efficient the conversion to ammonia. With improved control, pressure variations would be reduced allowing the operating set point to be moved closer to the maximum operating pressure. This increase in pressure and more constant pressure control will lead to increased production.

Another area of control recommended is control of primary furnace temperatures. The benefits of this control scheme are several fold. The digital system will prevent cycling and overfiring of the reform furnace. Either can waste fuel and decrease the life of the tubes and catalyst. The digital system will maintain tube gas temperatures within $\pm 5°F$ of the setpoint. Individual riser gas temperature set points can be maintained across the furnace. If the computer system is manipulating process gas to maintain loop H/N ratio, the furnace temperature control system will utilize a feedforward control scheme to maintain tight temperature control. Feedforward means that whenever gas is changed, fuel will be adjusted immediately to maintain temperatures before they change. In addition, feedforward is used to compensate for purge flow changes when purge is being burned in the furnace.

Maintaining furnace temperatures is important; however, methane leakage is the variable that must be controlled to increase production. At one plant, methane leakage was stabilized by simply maintaining constant temperatures. In other plants a methane analyzer is used in a feedback loop. The output of the methane analyzer is used to adjust temperatures of the furnace within limits to maintain methane leakage. In one plant methane leakage standard deviation was reduced an order of magnitude from $\pm.04\%$ to $\pm.004$.

Proper control of the synthesis converter bed temperature is another area of the plant that can benefit from digital control. The control scheme is based on an off-line program that calculates a quench flow distribution from data about the catalyst, bed size, catalyst area, etc. The on-line program maintains the quench distribution specified by the off line programs. The system also maintains bed one inlet temperature at a value specified by the model. Using this control scheme the operator will be able to attain the maximum per pass conversion from the synthesis converter. Temperature control is held to within $\pm 5°F$ regardless of converter total flow. The control scheme automatically adjusts for varying catalyst activity and converter flows. Normally quench flows are adjusted during start-up and forgotten; with the digital system the quench flows are continuously being adjusted for optimum temperature and flow in the converter.

A feature of the digital system that may be handled utilizing analog equipment is reform steam/gas ratio control. This control system is an inexpensive addition to the system. The purpose of steam/gas ratio control is to reduce steam usage by maintaining very tight ratio control usually to within $\pm.01$ ratio units. With good control of the ratio, the set point can be adjusted closer to the design value. An important advantage of the digital system is that before the steam valve is adjusted, all transmitters must be operating properly. This prevents control based on a bad transmitter. Both steam and process gas are pressure-temperature compensated so that a volumetric ratio can be maintained. In most cases significant steam savings are associated with good control of steam/gas ratio. It is a simple calculation to find the steam savings available in any plant. In several plant studies the savings has been estimated from 40,000 to 60,000 in steam. If the digital system is to manipulate process gas, during days the plant is air limited, to control loop H/N ratio, the steam/gas ratio control is a necessity to assure proper steam. In conclusion, steam/gas ratio is a very inexpensive addition to a digital control scheme that is justified by steam savings.

ADVANCED CONTROL SCHEMES

It is not enough to design and implement a sophisticated control scheme. As is true with a simple analog controller, the instrument is useless if the set point is not adjusted properly. The digital system must have proper set points also. Control of an ammonia plant can be divided into three levels. The first level of control is the conventional analog control supplied with every ammonia plant. This system requires that the operator make all decisions and set point adjustments. The second level of control and a degree more sophisticated is digital control of certain unit operations such as synthesis H/N ratio, loop pressure, methane leakage, and etc. This level of control interfaces to the first level and simply adjusts set points. For instance to control loop H/N, the second level of control would simply adjust air and gas flow set points. The second level of control consists of limited logic and a digital control algorithm. With two levels of control, the operator or process engineer must adjust the unit set points. It is important that these set points be reviewed at least weekly and, if possible, daily. To intelligently adjust the unit set points, the digital system must log or save operations data. The third and final level of control is the most sophisticated. It consists of relatively extensive logic and control schemes. Ideally the only operator inputs would be the philosophy as to how the plant is to operate. That is, maximum production or efficiency, limits on available gas, and any additional constraints peculiar to that days operation. The third level would then adjust set points in the second level to accomplish as best as possible the operator's request. This program must be custom designed for each user, however, some standard multi constraint programs are available currently.

The advantage of the third level of control is that of control enforcement. Once the optimum control scheme has been finalized, the digital system will perform repeatedly the proper control regardless of the operator on duty, training level, and etc. Digital control will assure that every day is your best operator's best day.

The justification for improved control is more production, more efficient operation, fewer shutdowns, and less maintenance. Not every ammonia plant can accurately measure product; however, in those that have conducted studies an increase of 1/2% to 3% has been documented. The variation in increases is due to the number of loops on control, how the plant operated before computer control and the attention given to the computer system. The digital systems installed to date have had a payout of 12 to 18 months. The digital system offers the capability to utilize available raw material supplies for the maximum profit and the opportunity to easily change the control scheme for varying market conditions.

ADDITIONAL BENEFITS

A further benefit of the digital control system is alarming and management reports. Each alarm condition is logged on a hard copy unit in the order in which the alarms occurred. Each message includes the time the alarm occurred, a description of the alarm, current value, and the engineering units. All process inputs and calculated values are checked for alarm conditions. Alarm limits are easily adjusted for changing conditions. Before an input is converted into engineering units and alarm checked, the input must be within a range indicating the transmitter has not failed. If a transmitter fails, a message is printed, the last good value is saved and the input is tagged as bad so that it will not be used for control.

The data logging capabilities can assist in determining the cause of alarms and recording past operating data. This data can be used to gain long-term improvements possible from plant studies and evaluations. Information from logging can be used to monitor for deterioration of equipment which might go unnoticed from operator's crude readings. In new plants a hardware savings is available by reducing the number of recorders in the plant and by adding vibration monitoring to the log-alarm system. Automated logging capabilities further relieve operators of time-consuming routine tasks -reducing the possibiliites for errors, elevating human responsibilities and opportunities for creativity, and yielding timely reports on schedule or on demand.

JUSTIFICATION

At this time demands for ammonia are at an all-time high at the same time the ammonia industry is facing serious constraints. In particular, natural gas, the primary raw material and energy source for ammonia production, is becoming increasingly expensive and in short supply. This presents an obvious challenge for the engineer to properly manage the available resources to gain maximum profit for the company. The digital system can make this job easier for the engineer. The digital system will operate the plant uniformly to eliminate shift to shift variations or fluctuations due to changing ambient conditions and operate closer to design margins to reduce waste of material and energy. Users indicate increases in production of 1/2 to 3% have been achieved. This makes computer applications in ammonia plants one of the leading application areas within the petrochemical industries.

AN IMPROVED COMPUTER CONTROLLED PROCESS CHROMATOGRAPH SYSTEM

Morton J. Hausner
Senior Associate Engineer
Mobil Research & Development Corporation
Princeton, New Jersey

ABSTRACT

Our Company has been successfully applying mini-computer based systems to the control of multiple process chromatographs in refineries since the mid sixties. At the end of 1974 it was decided to re-evaluate our system in the light of changing mini-computer and integrated circuit technology and the commercial availability of similar systems. The evaluation is discussed, leading to the conclusion that a new in-house system was the proper policy.

The original system had been conceived and designed to be used by chromatograph maintenance technicians, not programmers. The new system also follows this philosophy, but goes beyond the first system in its ease of man/machine communication.

The new system, which was designed, built and programmed by an outside systems house to our functional requirements, is described. Hardware and software are described and contrasted with the old system to show how size and cost were significantly reduced, while expanding the system's capability.

Interactive CRT displays are described to show ease of setting up chromatograph data tables, monitoring stream analyses, and recalling historical data.

System maintenance philosophy is described with particular emphasis on conversational diagnostics for every interface module provided by the system vendor.

INTRODUCTION

In 1974, after about eight years of successfully applying minicomputer-based chromatograph control systems in refineries, we arrived at a decision point. We had a reliable system built around a popular 12 bit minicomputer with paper tape as the program input medium and a teletypewriter as the man-machine communication device. The process interfaces to the chromatographs were put together from discrete modules involving considerable skilled labor to assemble and wire. The system filled three relay racks, one of which was devoted to wire terminations. Because of the high labor content, the cost of these systems escalated at a rapid rate. Moreover, the computer system did not lend itself to graceful expansion. Also, over a period of time, we had accumulated a list of desirable features to be incorporated into the system to improve its capability and ease of use.

At this point we had to decide whether to purchase a commercially available system or completely redesign our own system. After surveying commercially available systems, we concluded that a redesigned in-house system would provide the desired system capability at lowest cost.

We commissioned a local systems engineering firm to redesign, build and program the new system to our functional requirements.

SYSTEM DESCRIPTION

One design goal was to reduce system cost. This was accomplished by redesigning the process interfaces into plug-in printed circuit cards. This resulted in a reduction in system size from three racks to a single rack.

A second design goal was to make the system easier to program and to communicate with. We therefore chose a popular 16-bit minicomputer with a core-only real time operating system. Dual cassettes were used in lieu of paper tape. The man-machine communication device became a video display unit. An ASR teletypewriter was used as a hard copy device and back up communication device. Figure 1 shows a front view of the system.

INTERFACE HARDWARE

All process interface hardware takes the form of printed circuit cards which plug into an etched backplane which is actually a bi-directional bus connected to the computer's interface. The following types of cards are included in the process interface.

1. Bridge Control - Each bridge voltage signal is fed to its own multi-gain amplifier and analog to digital converter. No multiplexing is involved in chromatograph signal processing.

2. Valve Control - Digital output cards control up to ten relays for column and stream

valve switching.

3. Digital Inputs - 12 bit digital inputs are used to change from normal operation to calibrate mode.

4. Analog outputs permit specific components to be output to strip chart recorders for trending.

5. Miscellaneous Analog Inputs - The same card is used as for Bridge Control, with the addition of a four input multiplexer to the card. These cards are used to acquire data from other continuous and non-continuous process analyzers.

SOFTWARE

The system software has been written to be used by a maintenance technician having no knowledge of the program or programming in general. A set of alphanumeric CRT displays is made available to the technician for setting up the data tables that govern the chromatograph timing, switching and calculating functions. The technician need not concern himself with the location in memory of the various items of data, because memory is automatically allocated by the system software as it is needed. Each display is invoked by pushing a special function button. On each display, a menu of items is presented near the bottom of the screen. The technician selects an item he wishes to establish or modify by typing it at a specific location on the screen. Once an item is selected, the system guides the technician in providing the required information.

Of the nine available displays, three are used to allocate table space and fill tables with data.

1. G.C. Status

   This display permits the technician to allocate table space for the appropriate number of streams (one to five) for each G.C.. This display is the first one entered when initially setting up a G.C. Stream sequencing, noise factors, alphanumeric descriptors, time and date and other elective options (such as teletype output of analysis results and computer-to-computer transfer) are inserted through this display. Figure 2 shows this display with the menu of items at the bottom. The item chosen is typed in on the command line.

2. K-Factor, Elution Times, Limits

   For each stream every component is identified, its K-factor and elution time are assigned, along with a tolerance on elution time. The elution time is the time of occurrence of a peak maximum, and is obtained from a trial run of the chromarograph stream. If table space has been allocated in the G.C. status display for setting componential limits, the upper and lower limits of any or all components may be assigned. When a component violates its preassigned limits, the system can type out a message to that effect.

3. Time Function

   Based on the trial run of each chromatograph stream, various calculation mode changes and valve switching functions are assigned specific times of occurrence in this display. The menu of available functions is displayed at the bottom of the screen. These functions are easily changed and the system arranges these functions in the correct chronological order.

The above three displays take care of setting up G.C.s. A fourth display, called Analysis, permits the technician to view the latest analysis results of any stream. The technician may also assign any and all stream analysis results to be printed out on the teleprinter.

A fifth display, called Valve Control, permits the valves associated with any chromatograph (sample inject, stream valves, column valves) to be manually controlled or placed in automatic mode control of the normal program. This display is useful in initial system checkout or later debugging.

The above five displays are sufficient to do the basic job of chromatograph control for which the system was originally conceived. There are, however, four additional displays which take the system beyond its original functions.

1. Miscellaneous Analyzers

   The system is also capable of monitoring (not controlling) the operation of up to ten other continuous and/or discontinuous process analyzers. This display permits (1) the assignment of an analyzer to a given channel, (2) scaling of the input signal to provide proper engineering units, (3) the assignment of alphabetic descriptors, (4) sampling based on either a time interval (for continuous analyzers) or an interrupt from the field (for discontinuous analyzers).

2. Analog Output

   This display permits the assignment of any G.C. stream component or any miscellaneous analyzer signal to an analog output channel. The present intent for this function is to do trend recording for operator guidance. However, this feature lays the groundwork for future closed-loop control capability.

3. Cassette Control

   This display may be looked upon as a system utility which permits convenient saving of the entire contents of memory on a cassette and the reloading of the cassette contents back into core.

4. Histories

   This display permits the long-term retention of stream analyses, either total analyses or specific components. Total analyses are stored on cassette, while specific components are stored in core, space permitting.

ON-LINE WARNINGS

A number of features are available to detect impending trouble within the chromatograph:

1. Elution time tolerance - A time tolerance band is set for each component. If the elution time goes outside the band an error message is printed.

2. Area limit - A percentage tolerance is set around the total integrated area of the bridge signal. When this tolerance is exceeded, an error message is printed.

3. Missing peaks - A message is printed if a component expected at a certain time was not found.

4. Double peaks - A message is printed if two peaks occur within the same elution time tolerance.

COMMUNICATION WITH AN UPPER LEVEL PROCESS COMPUTER

A standard asynchronous serial link (RS232-C or current loop) with appropriate communication software is available to transmit analysis data to an upper level computer. Communication software must also be programmed in the upper level computer to check the data and ask for re-transmission if errors are detected.

MAINTENANCE PHILOSOPHY

The system consists of equipment from two sources. First, the central processor, cassette drives, and CRT display unit come from a mini-computer vendor with worldwide service and support capability. Although this hardware comes with a complete set of diagnostics, we recommend that our clients rely on the mini-computer vendor for maintenance and service support.

Second, the process input-output hardware comes from the system supplier. Here, we rely on a combination of thorough system training, good documentation and easy-to-use diagnostics for each interface module. When the system is ready for shipment to the plant site, appropriate personnel are brought into the system suppliers facility for training. The training covers both normal system use, plus the interface diagnostics.

The diagnostic programs were written for off-line set-up and testing of all interface cards. ISA Recomme

Recommended Practice RP55.1 was used as a guide in establishing the test procedures and expressing test results. The diagnostics are also presented in menu fashion and the technician is led through each test in a straightforward manner.

The diagnostics help to implement the philosophy of fault detection and card replacement. Spare cards are therefore normally included with every system. The final ingredient that contributes to a successful system is documentation. Four separate documents are supplied:

1. User Manual - This is the primary vehicle for running the system and is used as a guidebook during the training period.

2. Hardware Manual - This document consists primarily of detailed drawings, explanations of the various interface cards as well as parts lists.

3. Diagnostics Manual - This document explains how to use the interface diagnostics, which are stored on a cassette.

4. Software Manual - This document provides flow charts of all major program modules, plus program listings containing explanatory comments.

CONCLUSIONS

The redesign of this system has resulted in a three to one compression in size, about a 30% reduction in cost and a considerable increase in capability and growth potential. Since we are just beginning to accumulate field experience with this system, we cannot cite any long term up-time figures. We believe, however, that the redesign will result in a system that is more reliable, maintainable and powerful than the previous generation.

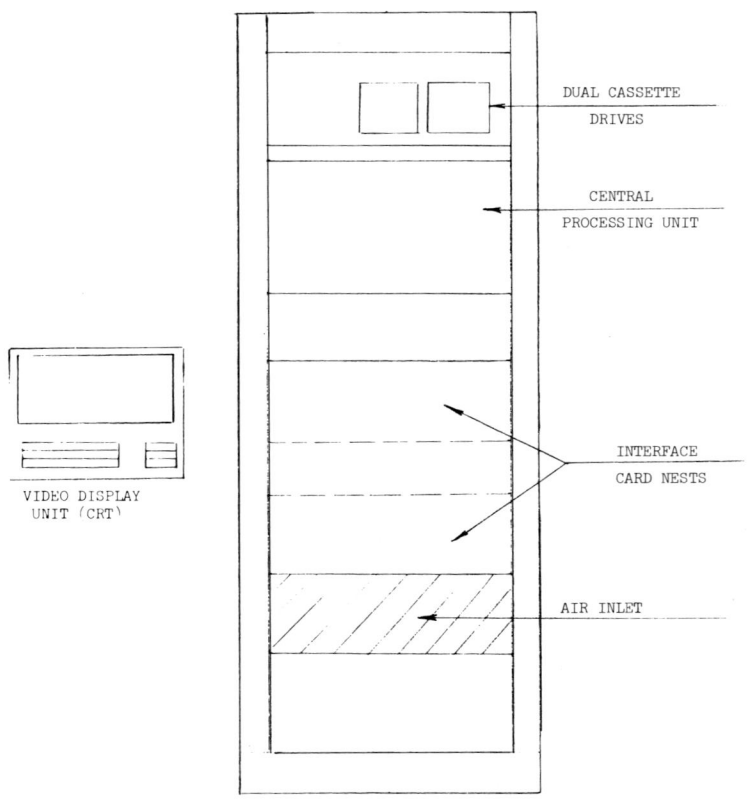

FRONT VIEW
MOBIL COMPUTER CONTROLLED CHROMATOGRAPH SYSTEM

FIGURE 1


```
GC STATUS -- OFF
                                GC = Ø1
STREAM   KF    FT    TF    LIMITS   TTY   CC    HYDR   AREA

Ø1  ON   TBL   TBL   TBL   A        ON    ON    ON     ON
Ø2  ON   TBL   TBL   TBL   A        ON    ON    OFF    ON
Ø3  OFF                    NA       OFF   OFF   OFF    OFF
Ø4  OFF                    NA       OFF   OFF   OFF    OFF
Ø5  OFF                    NA       OFF   OFF   OFF    OFF

EXPECTED PEAKS = 1Ø

NOISE VALUE = +5ØØ.Ø

STREAM SEQUENCE = 12345ØØØØØ


ALPHA DESCRIPTOR

Ø1   OVERHEAD PRODUCT
Ø2   BOTTOMS PRODUCT
Ø3
Ø4
Ø5

COMMAND =

GCS EXP NOISE SEQ ST KF FT TF ALPHA LIM HYDR AREA TTY CC TOD
    DATE
```

GAS CHROMATOGRAPH (G.C.) STATUS DISPLAY

FIGURE 2

©ISA, 1977
ISBN 87664-363-2

HIGH-SPEED DATA ACQUISITION OF EVENTS ON A VOLATILE PROCESS

George E. Pease, Jr.
Senior Engineer
Engineering Department
E. I. du Pont de Nemours & Co.

L. C. Boxhorn
Director of Marketing
I/C Engineering Corporation

ABSTRACT

Some volatile processes are subject to instability and often the basic analog controls are supplemented by critical event controls for protecting operating personnel and capital investment. A computer-based, high-speed system using remote multiplexing and direct memory access data acquisition provides a unique system for high resolution analysis of critical events involved in process shutdowns.

INTRODUCTION

Some volatile processes are subject to instability and are difficult to control. Generally, conventional electronic analog controls running the process are supplemented by a critical event control system. This sensitive system is to protect employees from potentially unstable process conditions and to protect substantial capital investment. When a shutdown occurs, an analysis of the shutdown cause is required, and complete corrective action must be taken before process start-up is authorized.

To reduce the time required for analysis and reduce the number of shutdowns, a high-speed computer-based data acquisition system was installed to monitor the critical event inputs and the process variables used in the shutdown analyses. The cost of the system was justified on reducing downtime. The feasibility of achieving this goal had been demonstrated on an older, slower system in use for several years.

DISCUSSION

The critical event system provides for analysis of:

- Analog data scanned at 1000 points/second with 12 bit bipolar A-to-D conversion accuracy.
- Digital inputs scanned at 120 points/second.
- Sequence of events (SOE) inputs scanned at 21,600 points/second.

The computer saves all this data, and, when an event occurs, it:

- Stores the SOE's in real time.
- Sorts the data for the process system involved.
- Saves a packet of data around the event.
- Prepares the data for presentation and analysis as hard copy or graphic CRT displays.

Remote Multiplexing

Remote multiplexing of these inputs was chosen because the system elements are geographically diverse (Figure I). The computer, multiplexer control and some display elements are located in a controlled environment near the plant technical offices 500 feet from the central control room (CCR). In the CCR are located six of the seven multiplexers and process displays and message typers. The seventh multiplexer is located 200 feet away from the CCR in the process area where it is subject to much vibration from the process equipment.

Remote multiplexing greatly reduced the cost of process wiring, especially the cost of 80 pairs of thermocouple extension wire run several hundred feet from the CCR and process to the computer room. Costs were further reduced by early shipment of multiplexer cabinets with prewired terminations allowing greater use of plant forces for signal hook-up before system delivery and start-up (Figure II).

Remote multiplexing does require very secure communications with sophisticated communication protocols and transmission techniques. Several systems are available, but the one used here required only two pairs of twisted, shielded wires for interrogation/commands and responses. Also, different modulation techniques are used on each: frequency shift keying (FSK) out and phase shift keying (PSK) in. Other protective measures include 1500 VAC isolation on every input, output and the data link, with bus isolators to protect the data path, and Geometric Coding for error detection. Additionally, the multiplexers are daisy chained and require only one set of the communications wires.

The nature of the process requires high reliability design of the multiplexing system elements. Enhancing the basic design is a burn-in requirement for the printed circuit boards and thorough factory testing of the assembled multiplexing system prior to shipment. Not all vendors were prepared to offer burn-in or factory testing.

Burn-in

System burn-in at high temperature (70°C) eliminates the majority of normal infant mortality. A secondary benefit is early detection and replacement of a bad lot of components which could cause field failures. Combined with extensive factory and field testing, factors like these enhance system availability. Obviously, to detect events and protect investment, the system must be operating properly.

Computer Interface

The computer normally must drive the multiplexer to acquire 400 analog inputs at 0.2, 0.5 and 1.0 second scan periods, collect 120 digital inputs on a 1.0 second period and accept SOE inputs to 5 millisecond resolution. In addition, a 64 port Scanivalve Pressure-to-Digital pneumatic scanner must be run with 0.5 and 1.0 second scan periods.

All this I/O could be done adequately to meet original project objectives using Programmed I/O (PIO). However, PIO was estimated to require 25% to 35% of compute time and left little expansion capacity (Figure III). Spending 25% of the compute time for the elementary task of data acquisition seemed wasteful and other data acquisition methods were explored.

After a great deal of investigation for feasibility of alternate I/O control methods, direct memory access (DMA) data acquisition seemed practical and efficient. DMA data acquisition was estimated to use 1% to 5% of compute time, depending on where the border between data acquisition and data handling is drawn. Using DMA, the system capacity is only limited by the maximum core data buffer size and the ability for intelligent handling of data for mass storage and process use. Additionally, more compute time for data analysis is available after a shutdown of one process system, while still collecting data on other process systems. The benefit is that less time is required to analyze 450 K words of data from each shutdown, and the process system may be restarted sooner.

Two basic system configurations offering DMA were designed for vendor proposals. One system is a hybrid using two computers (Figure IV A): one computer for scanning and one for data analysis. The scan computer is dedicated to data acquisition and is located remotely in the CCR. Communications to and from the main computer are over a full duplex, DMA data link requiring little main computer interaction.

The other system (Figure IV B) has a hardware controller for the remote multiplexers and DMA data transfer to the main computer. This system also provides interrupt for the SOE's and a programmed I/O port.

The PIO system and the two DMA systems could meet system functional requirements. The hardware controlled DMA system was chosen for these reasons:

- Only one computer to maintain.
- Simpler software in only one computer.
- SOE's come in on interrupt.
- A simple PIO interface.
- An established product.

Although the DMA interface is a new development, the controller and multiplexers are standard Uniplex 600 products manufactured by I/C Engineering Corporation.

Computer

The computer is a Digital Equipment Corporation PDP-11/70 with 250 K words of core memory and an 80 megabyte moving head disk for mass storage. The system has a floating point processor and DEC's Fortran IV Plus compiler operating under the RSX-11M real time executive. The floating point processor and Fortran IV Plus compiler are helpful because the software (except handlers) is all written in Fortran.

Software

The software system has four major packages running under the RSX-11M multiple task real-time operating system (Figure V). The first is the DMA data acquisition package which dirrects data alternately into one or the other of two 15 K word core data buffers. While one buffer is being filled, the other buffer is being moved into a 900 K word circular disk file. The DMA package also performs some housekeeping functions.

Next is a complete process monitoring and control package providing a fill-in-the-blanks data base, engineering units conversion, several types of alarms and CRT display of variables. The process package runs on a 60 second period and samples the DMA core buffer for process input data.

When an event occurs, an SOE interrupt is received through the PIO interface and the interrupt handler moves the SOE data and the time into a circular buffer included as a standard part of the RSX-11M executive. Then the SOE package forms a packet of data for the process system involved around the time of the first event. This packet is sorted from data for other process systems stored in the disk file and is put into an event disk file for use by the display package.

The display package formats the event disk file data and displays it on request on the graphics CRT terminal for analysis. The full packet of data is available in about 20 minutes. The display software was contracted to Biles & Associates and allows display of digital and analog data together on a common time scale (Figure VI). The time base may be expanded from 15 minutes to 20 seconds for resolution of events to 5 milliseconds.

The critical event control system described replaced an older system which provided service for several years for 100 points/second at 100 millisecond minimum resolution. Major improvement in process performance was accomplished with this system; however, the hardware is becoming obsolete and incentive exists to improve process performance.

CONCLUSION

As Figure V illustrates, the critical event control system is a unique, multitask system for high-speed data acquisition designed to take advantage of the strengths of each element in the system. The DMA data acquisition is under hardware control minimizing compute time required for this fixed, repetitive task. The majority of compute time is spent handling files and manipulating data to provide process surveillance and output data for human analysis. The remote multiplexing allows a high performance system with elements in many different locations. The event data and routine process data are presented on full graphics CRT's for easy interpretation by operating and maintenance personnel.

Using DMA for data acquisition is not a new technique. Some large, expensive systems of a few years ago used this technique, and some smaller laboratory systems today use DMA data acquisition for later off-line analysis. However, this technique is only now seeing use in the less expensive, smaller systems as discussed here. The unique advantages of real time process control and surveillance systems based on DMA data acquisition are gaining wider acceptance and, coupled with the flexibility of remote multiplexing, may become commonplace.

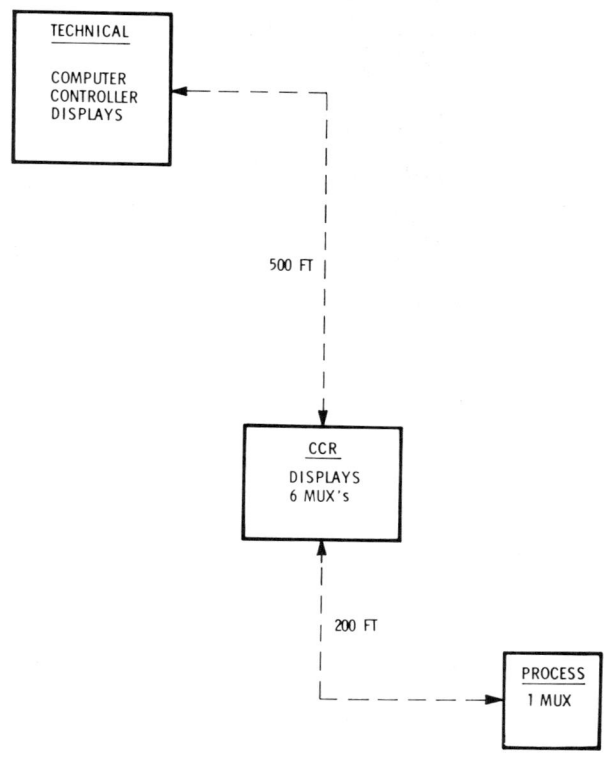

FIGURE 1

PREWIRED TERMINATIONS

FIELD WIRING

MULTIPLEXER CABLES

THIS PARTIALLY COMPLETE CABINET HAS TERMINATIONS FOR THREE REMOTE MULTI-PLEXERS.

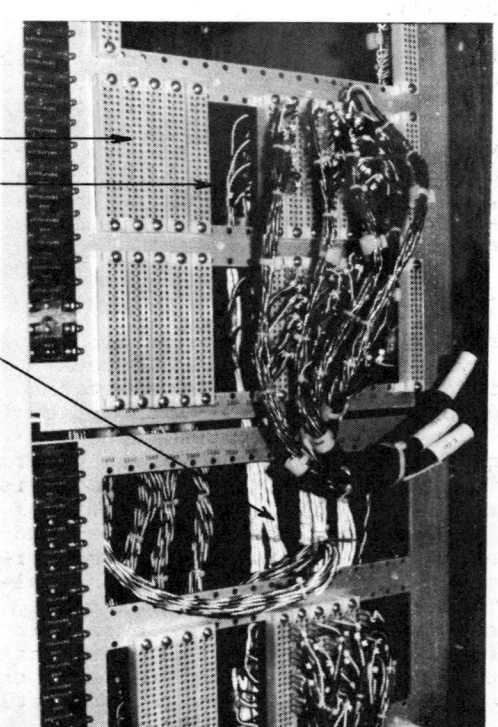

PIN TERMINALS

FIELD WIRING

MULTIPLEXER CABLES

PREWIRED PIN CONNECTORS WERE PROVIDED FOR ALL PROCESS INPUTS TO REDUCE TERMINATION SPACE AND WIRE HOOK-UP TIME.

FIGURE 2

PROGRAMMED I/O
REQUIRES 25% TO 35% COMPUTE TIME

DIRECT MEMORY ACCESS
1% TO 5% COMPUTE TIME

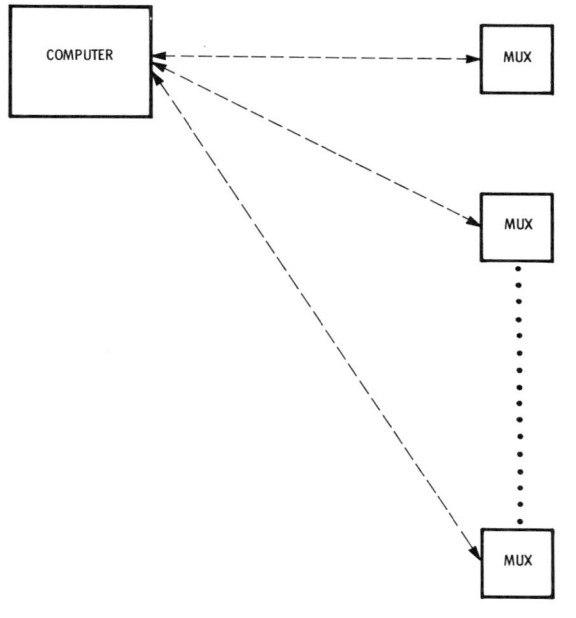

FIGURE 3

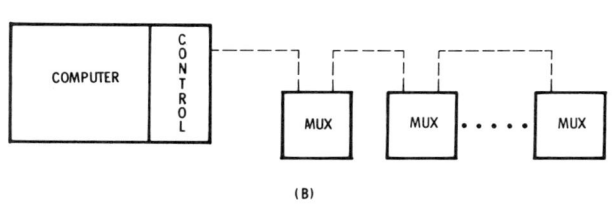

FIGURE 4

SOFTWARE

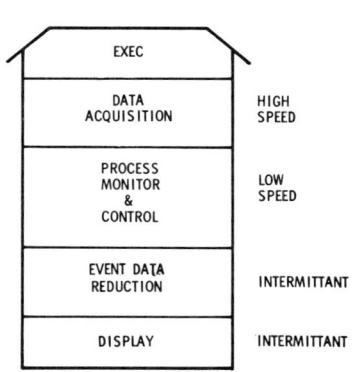

ALL SEGMENTS RUN CONCURRENTLY UNDER
A MULTIPLE TASK OPERATING SYSTEM

FIGURE 5

DATA DISPLAY

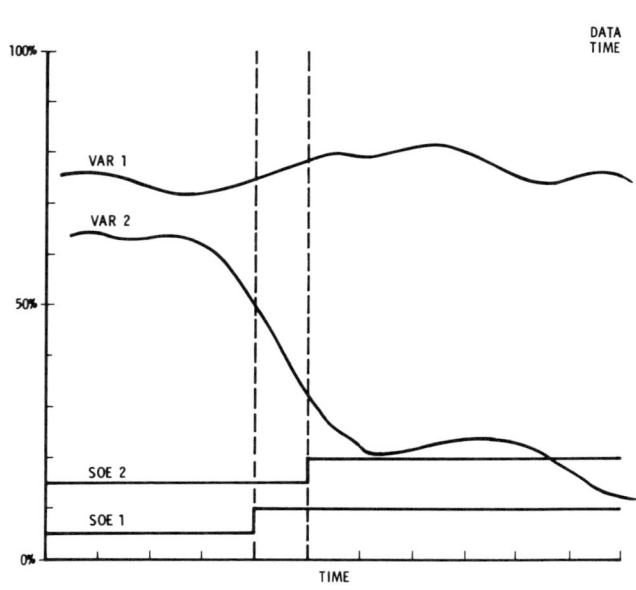

FIGURE 6

AUTHOR INDEX

L. C. Boxhorn 105
B. T. Condon 11, 21
A. B. Corripio 31, 57
T. P. Davis 47
T. F. Edgar 65
F. R. Groves, Jr. 37
J. W. Gruneisen 97
M. J. Hausner 101
J. O. Hougen 65
J. Martin, Jr. 31
C. F. Moore 77
G. E. Pease, Jr. 105
C. W. Ross 85
B. Sayers 77
C. O. Schwanke 65
C. A. Smith 11, 21, 37, 47
C. L. Smith 31
A. T. Touchstone 57
R. K. Wood 1

APPENDIX

The following papers were presented at the ISA/77 Anaheim Conference, but were not received in time to be included in the Proceedings:

NEW IDEAS IN PRACTICAL CONTROL SCHEMES, F. R. Groves, Jr.

ENERGY MANAGEMENT AN EVOLVING SCIENCE, T. M. Perkins, Jr.

ACKNOWLEDGMENTS
ISA NATIONAL OFFICERS 1976-1977

Division Participants

Director	K. L. Hopkins	Standard Oil Company of Ohio
Program Coordinator	R. C. Waggoner	University of Missouri—Rolla
Review Coordination	N. P. Brechtel	E. I. du Pont de Nemours & Co., Inc.

Session Developers

Session 1	R. C. Waggoner	University of Missouri—Rolla
Session 5	F. R. Groves, Jr.	Louisiana State University
Session 9	R. A. Mollenkamp	University of Missouri—Rolla
Session 12	D. M. Steelman	Leeds & Northrup Co.
Session 16	W. Pugh	Fluor Engineers & Constructors, Inc.

ISA/77 CONFERENCE COMMITTEE

Program Chairman	R. A. Kubick, Jr.	Ralph M. Parsons Co.
Assistant Program Chairman	S. Hammerstrom	C. F. Braun & Co.
Assistant Program Chairman	J. H. Jacoby	Ralph M. Parsons Co.
Assistant Program Chairman	J. R. Coffey	Rosemount, Inc.